FOOD SCIENCE AND TECHNOLOGY

TROPICAL FRUITS

FROM CULTIVATION TO CONSUMPTION AND HEALTH BENEFITS, PINEAPPLE

FOOD SCIENCE AND TECHNOLOGY

Additional books in this series can be found on Nova's website under the Series tab.

Additional e-books in this series can be found on Nova's website under the ebooks tab.

FOOD SCIENCE AND TECHNOLOGY

TROPICAL FRUITS

FROM CULTIVATION TO CONSUMPTION AND HEALTH BENEFITS, PINEAPPLE

CRISTINA STEWART BOGSAN

AND

SVETOSLAV DIMITROV TODOROV

EDITORS

nova
science publishers
New York

Library of Congress Cataloging-in-Publication Data

ISBN: 978-1-53612-885-7
Library of Congress Control Number: 2017959957

Published by Nova Science Publishers, Inc. † New York

CONTENTS

PREFACE

The pineapple is one of the most popular fruits from the Bromeliaceae family and is considered one of the most commercialized fruits in all continents. His popularity was grown since the time of the European explorers discovered this tropical fruit in the Americas, and they called them "pineapples". Pineapple is a fruit with different qualities and applications. Reach in vitamins and minerals, the fruit was appreciated by his nutritional value and applications in traditional medicine; modern medicine is use pineapple as a source for extraction of bromeline, a mixture of different proteolytic enzymes with application in treatment of numerous clinical disorders, including anti-inflammatory, anti-cancer, anti-microbial, coagulation, fibrinolysis and debridement agent; in culinary practice as a meat marinade and tenderizer; pleasant sweet-sour test is highly explored in preparations of desserts, drinks, ice creams, sweets, etc.

In the past pineapple was considered as a symbol of prosperity and been part of the iconography of the royalties and rich families in Europe. This was related to the fact that this fruit can be cultivated only in tropical areas of the World and in limited glass hothouses in Europe and North America. The pineapple was a symbol of a luxury. Like sugar, the pineapple was considered a commodity of privilege because the travel between Americas and Europe was expensive. In architecture, pineapple figures were considered as decorative elements, symbolizing hospitality and wellbeing. Related to the development of society, transportation and conservation techniques, nowadays we can explore all positives aspects of the pineapple. We can appreciate the nutritional qualities and can use as a source of extraction of bio-active molecules.

In this book we have to try to collect papers covering a different aspects of the systematic position of pineapple in the family of Bromeliaceae; agricultural aspects of pineapple production; studies of diseases related to this plant; genetic studies on melioration of the stability of the plant and its fruit and resistance to disease; nutritional profiles, health benefits, industrial applications; problems of collection, transportation and commercialization of pineapple; processing of the fruits and transformation to a nutritional products and extraction of bio-active compounds.

Cristina S. Bogsan and Svetoslav D. Todorov

In: Tropical Fruits
Editors: C. Stewart Bogsan et al.
ISBN: 978-1-53612-885-7

Chapter 1

Pineapple Taxonomy

Gabriel Moretti-Almeida*
Department of Biochemical and Pharmaceutical Technology,
São Paulo University, São Paulo, Brazil

Abstract

The pineapple (*Ananas comosus (L.) Merril*), Belongs to the family Bromeliaceae, which presents/ displays approximately 46 genera and 1700 species, occurring mainly in tropical zones.

The "pineapple" name are used by French, Italian, Dutch and German, being called by its form in Spanish "pineapple" and English "pineapple". Also, is employed in English and Portuguese (pineapple is the fruit, of course; Plant is pineapple), tropical fruit representation tropical and subtropical places, from America, was spreading around the world, principle these characteristics, smell and wonderful appearance. He adult pineapple has 70-80 leaves, with a length of 60-120 cm or more. The inflorescence of the pineapple is a closed spike, with numerous green or red bracts that cover the white or purple-white flowers.

The fruit consists in reality of 100-200 small units, of variable shape and size, the larger ones in the base, the smaller ones in the tip, in a set of conical shape. The conical fruit has a variable size, being able to accept

* Corresponding Author Email: gabrielmoretti@usp.br.

the average of 205mm of length, 145mm of diameter and 2200 grams of weight; The edible part of the fruit results from the thickened rachis which joins with the ovary pulp of the flowers. The pineapple finds eleven fruits production in the world.

Typical varieties Smooth Cayenne, Pérola, Queen, Singapore Spanish, Española Roja and Perolera. However, it is estimated that about 70% of world pineapple production comes from Smooth Cayenne. Pineapple is the eighth most produced fruit in the world, being the seventh one among those produced in Brazil. It is a fruit that has great importance for export, most of that in canned form than in fresh fruit, as can be observed in the painting.

INTRODUCTION

Pineapple is the eighth most produced fruit in the world, being the seventh one among those produced in Brazil. It is a fruit that has great importance for export, more in canned form than in fresh fruit (Loeillet 1995; Anon. 1998a). The pineapple, *Ananas comosus (L.) Mernil* belongs to the family Bromeliaceae, which presents about 46 genera and 1700 species, occurring mainly in tropical areas. Being called by its form in Spanish "pineapple" and English "pineapple," but the name "Ananas" used by French, Italian, Dutch and German, is also used in English and Portuguese (pineapple is the fruit, of course; Plant is a pineapple). The word pineapples originate from nana, from the Tupi language, spoken by the natives who inhabited the Brazilian coast. (Sanewski and Scott 2000). The root system of the adult plant is superficial, concentrating mainly on the first 15cm of the soil. The main stem, 20-30cm in length and 20-25mm diameter at its base, increasing up to 55-65mm at the top, is surrounded and covered by numerous leaves. Plants were originating from crown present the stem totally straight, (Collins 1960).

BOTANICAL CLASSIFICATION

The pineapple (*Ananas comosus L., Merril*) is a monocotyledonous plant, Herbaceous, of the family Bromeliaceae, (Kingdom: Plantae,

Division: Magnoliophyta, Class; Liliopsida, Order; Poales, Family; Bromeliaceae, Subfamily; Bromelioideae, Genus; Species: *A. Comosus*). Whose species can be divided about their habits, in two distinct groups: The epiphytes, which grow on other and terrestrial plants that grow in the soil At the expense of their roots. The pineapples Belong to the second group, more precisely to the genera *Ananas* and *Pseudananas*. Even presenting some characteristics epiphytes, such as the Capacity to store water both in the Special fabric of its leaves as in the Armpits of these (Kerns et al. 1936).

CHARACTERISTICS OF ANANAS

It originates from tropical and subtropical America, apparently from southern Brazil. Probably the present cultivated varieties are descended from existing wild pineapples. It is unknown, however, when, where and how this domestication took place, but on November 4, 1493, Columbus and his sailors discovered the pineapple in Guadeloupe in the Lesser Antilles (Linden 1879). Approximately 50 genera and 2.000 bromaliaceae species are known. Some species have ornamental value; others produce fine fibers for and manufacture of rustic material (cloth bag production), of fine fabrics, etc. Most Species is found under the Tropical regions of America conditions.

TAXONOMIC HISTORY

Since the first observations of pineapple by Europeans newcomers in the Americas to the present day, the taxonomy of pineapple varied considerably. The taxonomy of pineapple varied considerably, the first botanical description of the pineapple was carried out by Charles Plumier in the late seventeenth century (1755) when he collected plants calling them Karatas and pineapples on the island of Hispaniola. After the native classification, he created the genus Bromelia for the Karatas, in honor of the Swedish doctor Olaf Bromel, and described the pineapples, using polynomials like *Ananas aculeatus fructu ovato*, carne *alida*. In its *plantarum* species, Linnaeus (1753) designated the Pineapple as Bromelia pineapples and *Bromelia Comosa*, while Miller (1754, 1768) retained the name *Ananas*, with six varieties. In the following classifications of the eighteenth and nineteenth centuries, such as the Pineapple causing it to have numerous names for the same species or even genus (Leal et al. 1998).

Lindley (1827) used names such as *Ananassa sativa* for common cultivars, *Ananassa lucida* for smooth-leafed cultivars such as'mooth Cayenne (from the *Ananas (Lucidus*) variety of the eighth edition of miller's Gardener's Dictionary, published in 1768). *Ananassa debilis* for a peculiar cultivar where it is found corrugated leaves and *Ananassa bracteata* for a pineapple crowned with long bracts. Schultes and Schultes (1830) returned to the original name *Ananas*, with *A. sativus, A. debilis, A. semiserratus (instead of A. lucida) and A. semiserratus* Linden (1879) described a Colombian flat leaf cultivar with " Tubular" on *A. mordilona*.

Morren (1878) gave the first clear description of a deciduous pineapple, the Yvira, characterizing it by long bracts, absence of a crown and propagation by stolons which he called *A. macrodontes*.

In 1889, both Baker and Andre described a wild pineapple, with a long and small crowned fruit, respectively, under the names *Acanthostachys ananassoides* and *A. pancheanus*. Contrary to this amount of some, some authors like Bentham and Hooker (Andre 1889), claimed that the genus *Ananas* monospecific, with several wild species and cultivated in different ways. Thus only in 1892 published by Brasiliensis magazine, Mez recognized only one species, *Ananas sativus*, with five varieties of plants, a variety *Lucidos* includes pineapples with flat leaves and fruit, also a pinch or pitta, a small fruit only cultivated for fibers.

A. debilis was downgraded to a second variety *debilis*, 'only known from European glasshouses.' The variety *bracteatus* included *A. bracteatus* (Lindl.) Schult. f., but also *A. macrodontes* Morren. The variety *muricatus* was made from A. *muricatus* Schult. f., although Mez expressed doubts about its existence. The last variety, microstachys, corresponded to the wild pineapple *Acanthostachys ananassoides* Baker. In 1917, Merrill established the binomial *Ananas comosus*. In 1919, Hassler divided the genus *Ananas* in two sections *Euananas* and *Pseudoananas*. The latter was raised to a distinct monospecific genus by Harms (1930), with *Pseudananas macrodontes* (Morren) Harms. In his second classification, Mez (1934) did not recognize this new genus, and proposed three species: (i) *A. comosus*, including the wild pineapples and the cultivated forms *sativus* (spiny types), *lucidus* (smooth-leaved pineapples) and *debilis* (with doubts); (ii) *A. sagenaria*, corresponding to *A. bracteatus* (Lindl.) Schult. f.; and (iii) *A. macrodontes* Morren. Unfortunately, the second simple classification of Mez was no more successful than the first. From 1934 on, pineapple taxonomy was dominated by the views of L.B. Smith and F. Camargo. These authors divided the genus *Ananas*, renaming and multiplying the species in a long series of publications, without describing new variation (Camargo 1939, 1943, 1956; Smith 1939, 1961, 1962; Camargo and Smith 1968; Smith and Downs 1979).

The resulting over classification into two genera and nine species (Smith and Downs 1979) has been severely criticized on the basis of practicability, as well as inconsistency with avail- 22 G. Coppens d'Eeckenbrugge and F. Leal able data on reproductive behaviour and morphological, biochemical and molecular diversity (Leal 1990; Loison-Cabot 1992; Leal and Coppens d'Eeckenbrugge 1996; Coppens d'Eeckenbrugge et al. 1997; Leal et al. 1998).

The Bromeliaceae are identified by the stellate or scale-like multicellular hairs and the unusual conduplicate, spiral stigmas (Gilmartin and Brown 1987). As described in ethnobotanical inventory of fruits from America (Bromeliaceae - CGIAR):

> "The characterization of pinapple comprises a short stem, a rosette of narrow stiff leaves, terminal inflorescences in the form of racemes or panicles, hermaphroditic and actinomorphic trimerous flowers with well differentiated calyx and corolla, six stamens and superior to an inferior trilocular ovary, with axile placentation and numerous ovules. Fruits are capsules or berries and contain small naked, winged or plumose seeds,

with a reduced endosperm and a small embryo. Most species are epiphytic or saxicolous, but some are terrestrial. They are particularly adapted to water economy, based on: (i) rosette structure; (ii) ability to absorb water and nutrients through their waxy leaves and aerial roots; (iii) ability to store water in specialized aquiferous leaf tissue; (iv) multicellular trichomes functioning as water valvulae and reflecting radiation; (v) a thick cuticle; (vi) location of stomates in furrows, limiting evapotranspiration; and (vii) CAM. Their root system is not well developed and functions mostly to anchor the plant" (Bromeliaceae - CGIAR).

The *Bromeliaceae* are divided into three subfamilies, the *Pitcarnioideae*, the *Tillandsioideae* and the *Bromelioideae*. The attributes that characterized each family are several times independently as the adaptations to their environment. In the other hand, molecular studies change the subfamilies at the base of the *Bromeliaceae* in the contradictory form (Clark and Clegg 1990; Givnish et al. 1990; Ranker et al. 1990; Terry et al. 1997).

As an example, the *Pitcarnioideae* were long held to be the most archaic but their monophyly are questioned, and new subfamilies could be defined (Ranker et al. 1990; Terry et al. 1997) as the subfamilies *Tillandsioideae* and *Bromelioideae*. They are almost terrestrial, with armed leaf margins, hypogynous or epigynous flowers. The seeds are dry dehiscent capsules adapted to wind dispersal.

The *Bromelioideae* plants are mostly epiphytic, spiny, with epigynous flowers and fleshy or leathery berries containing naked seeds. All the species examined exhibit CAM, except those of the genus Greigia (Medina 1990). Pineapple evolved from eastern Brazil and the Amazon basin; the fruit had shown a tendency to the fusion of their carpels to make an indehiscent fruit, the sepals, petals, and filaments also are fusion. “Ananas capped the fusion tendency by merging the whole inflorescence, flowers, bracts and all into one massy compound fruit” (Smith 1934).

To the economic importance, Pineapple is the most important plant in the *Bromeliaceae* family. The fruit most consumed are under names like cardo or banana-do-mato (bush banana), piñuelas (small pineapple) or karatas, gravatá and croata (derived from Amerindian names given to terrestrial bromeliads), commonly consumed locally. Some bromeliads species are cultivated as ornamentals, used in traditional medicine or gathered for fiber extraction (Corrêa 1952; Purseglove 1972; Reitz 1983; Rios and Khan 1998 apud in Grovida 2017).

VARIETY CHARACTERISTICS

Main Varieties

The commercial production of pineapple is based on the Smooth Cayenne, Pearl, Queen, Singapore Spanish, Spanish Red, and Perolera.

However, it is estimated that about 70% of the world's Pineapple comes from Smooth Cayenne. At the Brazil and other Latin American countries, several local varieties and populations wild pineapple belonging to the Genus Ananas. Some of these materials could be recommended directly as varieties or used in genetic improvement, once characterized and properly evaluated.

The predominance of Smooth planting Cayenne in the main producing countries of the world; The use of few varieties for commercial plantations and the replacement of by Smooth Cayenne come causing the disappearance of varieties of local or regional interest.

Smooth Cayenne

Commonly known as pineapple Hawaiian, is the most planted variety in the world, both concerning area, and latitude, and is currently considered the queen of the varieties of pineapple because it has many characteristics favorable. It is a robust, semi-erect plant whose leaves do not have spines, other than some found at the apical edge of the leaf edge. The fruit is attractive, slightly cylindrical, weighs From 1.5 kg to 2.5 kg, showing yellow-orange peel when ripe, yellow flesh, rich in sugars (13° Brix at 19°

Brix) and of higher acidity than the other varieties. These characteristics make it suitable for industrialization and export as fresh fruit. The crown is relatively small, and the plant produces few puppy-like seedlings. In conditions of hot and humid climate, produces fruit Fragile for transportation and industrial processing. It is quite susceptible to withering Associated with the cochineal *Dysmicoccus brevipes* and the *fusariosis Fusarium subglutinans*. It was introduced in Brazil, in São Paulo, in the thirties, and later spread to other states, such as Paraíba, Minas Gerais, Espírito Santo, Goiás and Bahia, where from 1960 also assumed Increasing economic importance.

Singapore Spanish

It is the second variety in importance for industrialization, being largely cultivated in Malaysia because it is adapted to turfous soils of this and other South Asia. The plant presents medium size, with dark green leaves whose length Varies from 35 cm to 70 cm. Spin- Variable, occurring clones completely without spines and others with few thorns the edges of the leaves. The fruit is small, weighing 1.0 kg to 1.5 kg, cylindrical, with low Sugar (10° Brix - 12° Brix) and low acidity. The plant is vigorous, with Regular changes of the puppy and seed types. It is frequent the occurrence of crown Multiple. It has some resistance to Pests and diseases.

Queen

Variety widely cultivated in the South Africa and Australia. The plant is Small, with 60cm to 80cm of height, vigorous, with silvery leaves, small and with occurrence of dense spines. Produces a large number of bursts, but the number of pups is variable and are poorly developed.

The fruit is small (0.5kg to 1.0kg) with yellow bark, small, prominent eyes. The flesh is yellow and sweet (14° Brix to 16° Brix), low acid,

Excellent taste, and long life Post-harvest. This cultivar has to be explored in Brazil, due to Have some similar characteristics To the Pérola variety.

Spanish Red

Also known as Network Spanish, its plants are of medium size, vigorous, with dark green leaves Small spines and short spines. Thorny or partially prickly. Fruit of medium size (1.2kg to 2.0kg). In the form of a barrel, white or yellow- Pale, juicy, sweet-tasting. Total soluble solids around 12° Brix and low acidity, with a pleasant aroma.

Pearl

Broadly cultivated in Brazil, it is also known as Pernambuco or "Branco of Pernambuco". The plant has postage medium and erect growth; vigorous, with leaves about 65 cm in length and thorns on the edges. The peduncle of the Fruit is long, about 30 cm. Close to the base of the fruit, which Conical shape, yellowish bark (when Mature), white pulp. Produce soluble total solids from 14° Brix to 16° Brix, low acidity, pleasant to the palate of the Brazilian. The fruit weighs from 1.0kg to 1.5kg, has a large crown and has been little used for export. It presents wilt tolerance associated *Cochineal Dysmicoccus brevipes* and is susceptible to fusariosis, disease caused by the fungus *Fusarium subglutinans*.

Perolera

Variety planted commercially in Colombia and Venezuela, adapted to altitudes of up to 1500m. In studies performed by Embrapa, this cultivate behaved as resistant to fusariose. The plant has a height of 51.0cm, long peduncle, 29.2cm long, dark green leaf and maple, showing a little silvery band. Pronounced, one to two burst and eight to ten puppies. Fruit of form with an average weight of 1.8kg, of peel and pulp, with solids content total solids around 13° Brix, acidity titratable at about 10.0 meq/100 mL and high ascorbic acid content. In this cultivar tipping may occur of fruits, due to having peduncle long.

OTHER VARIETIES

Other varieties are planted in reduced scale for local markets and regional organizations, especially in Latin America. In Brazil, a variety of called Jupi, which is very similar to Pearl, from which it differs only by the cylindrical shape of the fruit, can be found In plantations in the Paraíba and Pernambuco. This variety, currently being disseminated in the Tocantins and Goiás. The predominance of the Pearl plantation and Smooth Cayenne in Brazil has caused the disappearance of planting varieties like Boituva and Rondon, once planted in various regions of the country.

Whatever the variety used, the farmer should be concerned about the maintenance of their characteristics morphological and agronomic. Although the Pineapple being a plant of propagation vegetative growth, continuous use of the same planting can provide the degeneracy of the clone by the accumulation of plagues and diseases, and the emergence of plants with characteristics different from the variety.

In the Smooth Cayenne variety, the appearance of plants with totally prickly leaves and fruit with fasked crown and, in Pearl, plants which produce fruits without a crown. Seedlings From these plants should be eliminated for the variety to maintain its Varietal standard.

SOIL

The perfect soil present mild to moderate acidity (pH from 4.5 to 5.5), medium texture (from 15% to 35% clay and more than 15% sand) and not subject to waterlogging.

Temperature

High temperatures during the day and low at night is the best to the fruit development. A variation around 8°C to 14°C between maximum and minimum temperatures, contributes to improve the quality of the fruit, mainly reducing its acidity, which is important for your consumption. Fruits produced during the warm month's present Lower acidity, as well as more pleasant aroma and flavor. The reverse occurs in cold periods.

The main characteristics desired in a variety of pineapples are (Coppens d'Eeckenbrugge et al. 1997):

- Growth fast;
- Short leaves, broad and without thorns;
- Early blight production located at the base of the plant ground;
- Production of offspring two centimeters from the base of the fruit;
- Fruit yellow-orange peel;
- Eyes flat, yellow flesh, firm but not fibrous;
- High sugar content, moderate acidity;
- Middle crown to small.

References

Anon. (1998) FAOSTAT Database. *Food and Agriculture Organization of the United Nations*. [Accessed 2000.] Available from http://apps.fao.org/.

Camargo, F. (1939) Ananás e abacaxí. *Revista de Agricultura de Piracicaba* [Ananás and pineapple. *Piracicaba Agriculture Journal]*, 14, 321–338.

Camargo, F. (1943) Vida e utilidade das Bromeliáceas. Boletim Técnico Instituto Agronômico do Norte [Life and utility of Bromeliads. *Technical Bulletin Agronomic North Institute]*, Belem, Pará, 31 pp.

Camargo, F. (1956) Nota previa. Ananas lyman-smithii n. sp. *Arquivos do Jardim Botánico do Rio de Janeiro* [Previous note. Ananas lyman-smithii n. sp. *Archives of the Botanical Garden of Rio de Janeiro]*, 14, 281–289.

Camargo, F. and Smith, L. B. (1968) A new species of Ananas from Venezuela. *Phytologia,* 16, 464–465.

Clark, L. G. and Clegg, M. T. (1990) Phylogenetic comparisons among rbcL sequences in the Bromeliaceae. *American Journal of Botany*, 77, 115.

Collins, J. L. (1960) *The Pineapple, Botany, Utilisation, Cultivation*. Leonard Hill, London, 294 pp.

Coppens d'Eeckenbrugge, G., Leal, F. and Duval, M. F. (1997) Germplasm resources of pineapple. *Horticultural Reviews*, 21, 133–175.

Corrêa, M. P. (1952) Diccionário das plantas úteis do Brasil e das exóticas cultivadas, Vol. III. Imprensa Nacional, Rio de Janeiro. [*Dictionaries of useful plants in Brazil and cultivated exotic species, Vol. III*. National Press, Rio de Janeiro].

Gilmartin, A. J. and Brown, G. K. (1987) Bromeliales, related monocots, and resolution of relationships among Bromeliaceae subfamilies. *Systematic Botany*, 12, 493–500.

Givnish, T. J., Sytsma, K. J. and Smith, J. F. (1990) A re-examination of phylogenetic relationships among bromeliad subfamiles using cpDNA restriction site variation. *American Journal of Botany*, 77, 133.

Harms, H. (1930) Pflanzenfamilien. Bromeliaceae. *Engler Prantl*., 65–159.

Kerns, K. R., Collins, J. L. and Kim, H. (1936) Developmental studies of the pineapple Ananas comosus (L) Merr. *The New Phytologist*, 35, 305–317.

Leal, F. (1990) Complemento a la clave para la identificación de las variedades comerciales de piña Ananas comosus (L.) Merrill. *Revista de la Facultad de Agronomía (Maracay),* 16, 1–11. [Complement to the key for the identification of commercial pineapple varieties *Ananas comosus* (L.) Merrill. *Journal of the Faculty of Agronomy* (Maracay)].

Leal, F. and Coppens d'Eeckenbrugge, G. (1996) Pineapple. In: Janick, J. and Moore, J. N. (eds) *Fruit Breeding*. John Wiley & Sons, New York, pp. 565–606.

Leal, F., Coppens d'Eeckenbrugge, G. and Holst, B. K. (1998) Taxonomy of the genera Ananas and Pseudananas – a historical review. *Selbyana*, 19, 227–235.

Leal, F., Coppens d'Eeckenbrugge, G. and Holst, B. K. (1998) Taxonomy of the genera Ananas and Pseudananas – a historical review. *Selbyana*, 19, 227–235.

Linden, M. J. (1879) Ananas Mordilona Linden. *Belgique Horticole*, 29, 302–303.

Lindley, J. J. (1827) Billbergia. *Botanical Register* 13, 1068.

Linnaeus, C. (1753) *Species plantarum.* Stockholm, Sweden, 724 pp.

Loeillet, D. (1994) The world concentrated pineapple juice market: has the bandwagon become an applecart? *Fruitrop,* 9, 10–16.

Loison-Cabot, C. (1992) Origin, phylogeny and evolution of pineapple species. *Fruits*, 47, 25–32.

Medina, E. (1990) Eco-fisiología y evolución de las Bromeliaceae. Boletín de la Academia Nacional de Ciencias [Eco-physiology and evolution of Bromeliaceae. *Bulletin of the National Academy of Sciences*], *Córdoba*, 59, 71–100.

Morren, E. (1878) Description de l'Ananas macrodontes. sp. nov. Ananas à fortes épines. *Belgique Horticole (Liège),* 28, 140–172. [Description of Pineapple macrodontes. sp. nov. Pineapple with strong thorns. *Horticultural Belgium* (Liège)].

Purseglove, J. W. (1972) *Tropical Crops*. Monocotyledons. Longman, London, pp. 75–91.

Ranker, T. A., Soltis, D. E., Soltis, P. S. and Gilmartin, A. J. (1990) Subfamilial phylogenetic relationships of the Bromeliaceae: evidence from chloroplast DNA restriction site variation. *Systematic Botany*, 15, 425–434.

Reitz, R. (1983) Bromeliáceas e a malária – Bromélia endêmica. Flora Ilustrada Catarinense, Santa Catarina. [Bromeliads and malaria -

Endemic Bromeliads. *Flora Illustrated Catarinense Santa Catarina*], 808 pp.

Rios, R. and Khan, B. (1998) List of ethnobotanical uses of Bromeliaceae. *Journal of the Bromeliad Society,* 48, 75–87.

Sanewski, G. and Scott, C. (2000) The Australian pineapple industry. In: Subhadrabandhu, S. and Chairidchai, P. (eds) *Proceedings of the Third International Pineapple Symposium. International Society for Horticultural Science*, Pattaya, Thailand, pp. 53–55.

Schultes, J. A. and Schultes, J. H. (1830) Ananas sativus n.n. *Systema Vegetabilis*, 7, 1283–1285.

Smith, L. B. (1934) Geographical evidence on the lines of evolution in the Bromeliaceae. *Botanische Jahrbücher*, 66, 446–448.

Smith, L. B. (1939) Notes on the taxonomy of Ananas and Pseudananas. *Botanical Museum Leaflet, Harvard,* 7, 73–81.

Smith, L. B. (1961) Notes on Bromeliaceae. *Phytologia*, 8, 12.

Smith, L. B. (1962) A new look at the species of pineapple. *Bromeliad Society Bulletin*, 12, 54–55.

Smith, L. B. and Downs, R. J. (1979) Bromelioidées (Bromeliaceae). Monograph 14, pt. 3, *Flora Neotropica*, The New York Botanical Gardens, New York, 2142 pp.

Terry, R. G., Brown, G. K. and Olmstead, R. G. (1997) Examination of subfamilial phyogeny in Bromeliaceae using comparative sequencing of the plastid locus NDHF. *American Journal of Botany*, 84, 664–670.

In: Tropical Fruits
Editors: C. Stewart Bogsan et al.
ISBN: 978-1-53612-885-7

Chapter 2

NUTRITIONAL VALUES AND POST-HARVEST HANDLING OF PINEAPPLES

Balakrishnan Kunasundari[1], Mei Kying Ong[2], Kokila Thiagarajah[3], Shao Yin Ooi[2] and Huey-Shi Lye[2,*]

[1]Faculty of Engineering Technology, Universiti Malaysia Perlis (UniMAP), D/A Pejabat Pos Besar Kangar, Perlis, Malaysia

[2]Department of Agricultural and Food Science, Faculty of Science, Universiti Tunku Abdul Rahman, Jalan Universiti, Bandar Barat, Kampar, Perak

[3]Department of Biomedical Science, Faculty of Science, Universiti Tunku Abdul Rahman, Jalan Universiti, Bandar Barat, Kampar, Perak

ABSTRACT

Pineapple *(Ananas comosus)* is a popular tropical and sub-tropical fruit that cultivated in more than 90 countries. It is a non-climacteric fruit with high nutritional value that can be consumed in fresh or processed form. Besides forming great sources of carbohydrate, fibre and minerals,

* Corresponding Author: Huey-Shi Lye, Email: lyehs@utar.edu.my.

it is also a promising dietary supplement with antioxidant properties. A number of studies found that bromelain, a glycoprotein that presents abundantly in pineapple contributes to numerous health benefits such as anti-inflammatory, anticancer, antimicrobial, coagulation, fibrinolysis and debridement. Pineapple has also been widely used in several manufacturing industries includes chemicals, food and animal feed. Post-harvest diseases, physiological disorders and poor handling are the major problems that lead to the post-harvest loss. Therefore, proper storage, packaging methods and efficient post-harvest treatments are practiced to prolong the shelf life and maintain the quality of the fruit. The present chapter deals with nutritional profiles, health benefits, industrial applications, post-harvest losses and handling, as well as safety assessment of pineapple.

1. Introduction

Pineapple *(Ananas comosus)* is a popular tropical and sub-tropical fruit. The crop, which is originated from South America now thrives in more than 90 countries. Sixty eight percent increase of global pineapple production was recorded in 2014 (25.43 million tonnes) when compared to year 2000 (15.11 million tonnes). Asia has been the prominent pineapple producing region, accounting for 43.9% of total production followed by Africa (18.1%) and South America (17.6%). The world's three largest producing nations, namely Costa Rica (2.92 million tonnes), Brazil (2.65 million tonnes) and Philippines (2.51 million tonnes) have been responsible for roughly 31.7% total production of this crop in 2014. Thailand (7.5%), Indonesia (7.2%), India (6.8%), Nigeria (5.7%) and China (5.6%) are the other leading countries in the global pineapple production (FAOSTAT, 2017).

Pineapple is a non-climacteric fruit which belongs to family Bromiliaceae and subfamily Bromelioideae. There are 56 genera with 2794 species associated with this family (Bartholomew et al., 2003). Pineapple species can be categorized based on the growth habit into two groups which are terrestrial and epiphytic. Typically, this crop belongs to a terrestrial group while still exhibiting some properties of epiphytic such as an ability to store a small amount of water in their leaf axils to survive dry season (Sharma et al., 2009). Pineapple is a herbaceous perennial plant that can attain a height of 1.0 to 2.0 m. Members of this family are characterized by short stems and sword shaped leaves with needle-like tips. The leaves are arranged spirally and tightly clasping a central stem. The color of the leaf varies from light or dark green to dark red or purple depending on the species and growth conditions. The fruits are cylindrical with short pointed arranged leaves referred as 'crown' at its apex (Bartholomew et al., 2003; Mohammed, 2004). Generally, the fruits are seedless with pale to yellow golden flesh and white at the core. As pineapple is a monocot, the flowers parts occur in multiple of three's, sepal three, short and fleshy three pale violet petals which form a tube and enclose 6 stamens, 1 style and 3 lobed stigma (Sharma et al., 2009). Each pineapple plant yields a single fruit prior to suckers generations that will grow into new plants (Mohammed, 2004).

There are number of pineapple varieties but the common ones are grouped into 'Spanish', 'Queen', 'Cayenne', 'Abacaxi' and 'Maipure'. These categorizations are based on fruit color, shape, flat or sharpness of 'eyes' and spine as well as the properties of the leaves as standard botanical or horticultural cultivar classification for pineapple. 'Smooth Cayenne' and 'Red Spanish' are commercially important varieties. 'Smooth Cayenne' is the predominant variety adapted for canning and processing. The fruit is slightly acidic, weighs 2-2.5 kg with pale yellow to golden yellow fruit pulp and possesses high sugar content. The leaves are smooth with spines found mostly on the tip. On the other hand, 'Red Spanish' flesh is pale yellow to whitish while acidic in taste and the fruit weighs about 1-2 kg. This variety tends to have very spiny leaves (Bartholomew et al., 2003; Mohammed, 2004; Sharma et al., 2009).

There are three distinct phases of pineapple growth which are the vegetative phase of leaf growth, the generative phase of fruit growth and another vegetative phase of shoot growth. It takes two to two and a half years to complete all of the phases. The fruit weight gain has been observed in a continuous sigmoid trend after initiation of inflorescence (Mohammed, 2004). Pineapple fruit eating quality is determined by sugar content and fruit acidity. High level of acidity is found in the initial fruit growth phase and reduces during fruit ripening. On the contrary, a high sugar content is observed when the fruit approaches maturity (Bartholomew et al., 2003; Saradhuldhat and Paull, 2007).

2. Nutritional Profile and Health Benefits

2.1. Nutritional Values

Pineapple has been consumed predominantly in fresh or processed form. Fresh consumption involves whole fruit or fresh-cut form that undergoes minimal processing with limited shelf-life. Chunks of fresh pineapples are added with some spices and sauces for salad preparation as well as in preparing pies or puddings. The processed products include canned pineapple, juice and dried forms that can be added into bakery items.

Pineapple is a good source of carbohydrate, minerals and fibre with low amounts of sodium and fat. The nutritional composition may vary based on the varieties, growing conditions and degree of the ripeness (Hassan and Othman, 2011). It has been reported that the nutritional value of pineapple is 50 kcal per 100 g fresh fruit (Table 1) (USDA, 2017). Glucose, sucrose and fructose are the major component of carbohydrates in the pineapple. High concentrations of glucose and fructose are found during fruit development stage while the accumulation of sucrose is detected during ripening stage. The presence of substantial amount of dietary fibre promotes the consumption of pineapple to ensure normal bowel movements. It also contains minerals such as calcium (Ca), potassium (K), magnesium (Mg) and phosphorus (P). Pineapple can contribute to the significant amount of daily manganese (Mn) intake as it is readily available in an assimilable form (Beattie and Quoc, 2000; Hassan and Othman, 2011). Apart from that, various types of B-complex vitamins such as thiamine, riboflavin and niacin are present in this fruit.

Table 1. Nutrient composition of raw pineapple (USDA, 2017)

Nutrient*	Unit	Pineapple
Water	g	86
Energy	kcal	50
Protein	g	0.54
Carbohydrate	g	13.12
Fiber	g	1.4
Vitamins		
Thiamin	mg	0.079
Riboflavin	mg	0.032
Niacin	mg	0.500
Vitamin A	IU	58
Vitamin C	mg	47.8
Vitamin E	mg	0.02
Minerals		
Calcium	mg	13
Iron, Fe	mg	0.29
Magnesium, Mg	mg	12
Phosphorus, P	mg	8
Potassium, K	mg	109
Sodium, Na	mg	1
Manganese, Mn	mg	0.927

* Value per 100g.

The ascorbic acid content is approximately 47.8 mg per 100 g fresh fruit during harvesting. There is a decline in ascorbic acid concentration following the maturity stage of the fruit (USDA, 2017). Pineapple is also a promising dietary supplement consists of antioxidant compounds such as ρ-coumaric acid, ferulic acid, caffeic acid, sinapic acid, ρ-coumaroyl quinic acid, ρ-hydroxybenzoic acid and ρ-hydroxybenzoic aldehyde (Hassan and Othman, 2011). These dominant antioxidants can exhibit scavenging activity on reactive oxygen species and free radicals (Prior and Cao, 2000). Additionally, investigations revealed that more than two hundred volatile compounds associated with pineapple aroma are mostly esters, lactones, hydrocarbons, aldehydes, alcohols and carbonyl acids. According to Brat et al. (2004), the pineapple flavor is majorly contributed by two thioesters which are methyl 3-(methylthio) propanoate and ethyl 3-(methylthio) propanoate. Meanwhile, citric and malic acids are the main organic acids detected with ripe pineapple fruit. Apart from that, bromelain, a glycoprotein having protease activity is found in appreciable quantities in pineapple. There are several therapeutic benefits associated with bromelain, such as reversible inhibition of platelet aggregation, bronchitis, angina pectoris, sinusitis and enhanced absorption of drugs, particularly of antibiotics (Maurer, 2001).

2.2. Medicinal Values

Pineapple has been used in traditional medicine practices such as in Ayurvedic, Unani, Siddha and so on. The practitioners have used it as anti-inflammatory, digester, an inhibitor of blood platelet aggregation and many more (Khare, 2008).

2.2.1. Anti-Inflammatory

Bromelain is a type of proteases that can be extracted abundantly from pineapple stem. Bromelain has been shown to have medical benefits in many inflammatory or immune mediated diseases. In an *in vivo* study, the researchers revealed that bromelain helps to alleviate pathological changes of asthma in female C57BL/6J mice. Half of the mice were induced with ovalbumin (OVA) which can lead to acute allergic airway (AAD) disease including asthma causing inflammation in the lung due to hyper responsiveness. Oral bromelain treated mice showed significant ($P<0.05$) reduction of airway reactivity, methacholine sensitivity, CD8+ and CD19+ lymphocytes, eosinophil count and cytokine level such as IL-13 when compared to the control group of AAD. Reduction of these inflammation biochemical markers was also supported by histological findings (Secor et al., 2008).

Pineapple juices can also be used as an adjuvant or supplement to alleviate other inflammatory diseases such as arthritis (Majeed and Barole, 2015). In another independent study, bromelain was found to reduce receptors for IL-8 induced neutrophil attachment and thus reducing the migration to the site of inflammatory stimuli which subsequently reduce the inflammatory response (Fitzhugh et al., 2008).

Apart from this, surgical removal of impacted teeth involves intentional trauma to the bone and soft-tissues that lead to inflammatory reaction and steroids are usually given to manage the pain and swelling (Saha, 2016). A recent study conducted on oral surgery involving third molar removal showed that 70% out of 40 patients treated with bromelain experienced less pain. Therefore, bromelain was suggested to be used as oral enzyme therapy to replace or reduce the steroids dosage (Singh et al., 2016). On the other hand, bromelain is recommended at the dosage of 500–1000 mg three times a day as an effective treatment option for the inflammatory sequelae of chikungunya virus (CHIKV) (Adrover-López et al., 2016).

2.2.2. Anticancer

Pineapple has also been studied for its anticancer properties. A list of gastrointestinal cancer cell lines of different origins, chemosensitivities and phenotypes such as MKN45, KATO-III, HT29-5M21, HT29-5F12 and LS174T were tested against mixture of bromelain and N-acetylcysteine (NAC) which is naturally found in *Allium* plants. The combination of these compounds was found to significantly ($P < 0.05$) suppress the cell proliferation by inducing cell death via both apoptosis and autophagy (Amini et al., 2014). These findings have also been proved in an *in vivo* study as the combination treatment caused the number of tumor nodules and tumor burden significantly reduced in mice with gastric or colorectal cancer. Moreover, no toxicity was observed (Amini et al., 2015).

Mostly, testing using bromelain showed promising results on ovarian, breast and gastrointestinal cancer cell lines (Pauzi et al., 2016; Amini et al., 2016; Gani et al., 2015). *In vitro* cancer studies on ovarian (A2780) and colon (HT29) cancer cell lines also showed that bromelain was able to suppress the growth and colony formation via apoptosis without affecting the normal cells (Gani et al., 2015), which is one of the basic requirement to develop a potent chemotherapy.

Numerous studies have been conducted to evaluate the relationship of bromelain and its anticancer activities. A few established mechanisms which make bromelain to be act as a potential anticancer therapy are inhibition of tumor cell growth and metastasis, regulation of inflammatory mediators, immune-modulatory activity, stimulation of neutrophils alteration of tumor micro-environment, regulation of hemostatic system and reduction of blood coagulation capacity. However, more clinical studies or clinical evidences are required to develop bromelain based chemoprevention and adjuvant cancer therapy (Chobotova, Vernallis and Majid, 2010).

2.2.3. Antimicrobial

Besides anti-inflammatory and anticancer activities, pineapple also possesses antimicrobial properties. Bromelain has been shown to have antimicrobial potential against *Escherichia coli*, *Enterococcus fecalis*, *Aggregatibacter actinomycetemcomitans*, *Streptococcus mutans* and *Porphyromonas gingivalis* (Praveen et al., 2014). Other than direct antimicrobial activity, bromelain may increase the absorption of antibiotics which subsequently enhance the antimicrobial activity of antibiotics (O'Mahony, 2010).

Pineapple has also been used to treat intestinal worm infection. In a clinical trial, fresh pineapple juices were given to 90 children who were positive for soil-transmitted helminthiases such as *Ascaris lumbricoides*, *Trichuris trichuria*, and other helminths. They were equally divided and treated with pineapple puree and mebendazole, a standard antihelminth drug. Pineapple puree treated group had a notable egg reduction rate of 83.52% and cure rate of 68.9% and this was closely resembled mebendazole treated group results which were 92.5% and 88.9%, respectively. Moreover, both interventions were well tolerated and no adverse effect was reported (Manabo and Fries, 2010).

2.2.4. Coagulation, Fibrinolysis and Debridement

Bromelain induces the synthesis of fibrin which plays an important role in clotting. At the same time, it also helps the succeeding process which is fibrinolysis, breaking down the formed fibrin or clot. This is achieved by increasing the conversion of plasminogen to plasmin which activates the fibrin breakdown (Pavan, Jain and Kumar, 2012). Debridement is the removal of damaged or dead tissue or foreign objects from a wound. Escharase found in bromelain is responsible for the debridement activity. In an *in vivo* study, a quick enzymatic debridement with a bromelain-based agent (Debriding Gel Dressing, DGD) has been studied on a pig. Quick dissolution of the burn eschar was noticed in all DGD treated tissue compared to control tissue which was only treated with a hydrating vehicle. DGD also showed no adverse effect on normal skin area which can make it as a potential enzymatic debriding agent (Rosenberg et al., 2012).

2.3. Industrial Applications

In the search for green or environmental friendly products, pineapple leaves, stem and crown can be used as fiber sources. The fiber is about 60cm long, white, creamy and lustrous. It easily takes and retains dyes (Malézieux et al., 2003). Recently, pineapple fibers have been widely studied to develop better thermoplastic composites such as polyester and polypropylene (Chollakup et al., 2011; Neto et al., 2013). The leaves can also be used as animal feeds (Joy, 2010).

Additionally, pineapple waste can be utilized to produce vinegar and vanillin. It also acts as an alternative source of energy (Lobo and Paull, 2016). A study revealed that pineapple waste was one of the most promising potential biogas producer compared to other wastes such as rice straw, palm oil waste and so on. Pineapple peel is recorded as the highest maximum specific methane production rate (36.77 ml/day) (Paepatung, Nopharatana and Songkasiri, 2009).

In the clinical laboratory, chemical anticoagulants have been used to prevent blood sample from clotting prior to any analysis for a diagnostic procedure to be performed. However, these anticoagulants such as ethylenediaminetetraacetic acid (EDTA) are expensive and toxic to human health. As an alternative, pineapple or bromelain can be used to prevent blood coagulation. A study has been conducted to compare the anticoagulation property of pineapple extracts and EDTA. Results revealed that pineapple extracts successfully inhibited blood coagulation for about three hours and produced similar results with EDTA in terms of preventing the hemolysis, red blood cell clumping, crenation, and maintaining cell size and shape. The study showed that pineapple extract can be a potential substitute for a standard anticoagulant (Pieris, Jansz, and Dharmadasa, 2014).

In addition, bromelain, the protease enzyme can be used to tenderize meat. A study has been conducted to determine the effect of bromelain on different meats such as beef, chicken and squid. Results showed that bromelain was able to reduce the firmness and toughness of the meat. This was due to the presence of plant thiol proteases in bromelain. Thiol proteases have very broad specificities in which it can breakdown the wide range of muscle proteins. Thus, it can be a good alternative for the chemical tenderizers (Ketnawa and Rawdkuen, 2011). On the other hand, bromelain can be used in the leather tanning process and stabilized latex paints (Joy, 2010).

3. Post-Harvest Losses and Handling

The harvest maturity of pineapple fruit is based on peel color and shape of individual fruitlets or eyes. Factors of season, rainfall, altitude and field practices may cause variation of the peel color of pineapple. The fruit is ready to be harvested when 30-50% eyes turned yellow from the base. The fruit is best distributed for local market and distant market when fruit is at 10-20% yellow stage and 10% green, respectively (Adikaram and Abayasekara, 2012).

Pineapple is a highly perishable food and continuously respiring even after harvest, thus limiting its shelf life and facing storage problem. While in storage, fruit is easily losing moisture due to transpiration and increased susceptibility to spoilage microorganism (Hossain and Bepary, 2015).

Hence, there is a need for proper post-harvest handling and management of pineapple.

In pineapple, fruit physical injuries, sun burn, rodent's infestation, post-harvest diseases, contamination with pathogenic fungi and bacteria are major causes of losses. Lack of awareness, knowledge and skills related to pre- and post-harvest management of produce among the handlers and marketers, high temperature condition and unavailability of efficient cool chain infrastructure worsen the post-harvest losses (Hossain and Bepary, 2015). The poor handling during transportation and use of inappropriate marketing structures also contribute to the loss.

Thus, reduction of post-harvest food losses is vital in ensuring the future global food security. Post-harvest loss of pineapple can be minimized by implementing systematic and proper planning of agricultural practices on top of increasing the standard of various post-harvest handling practices.

3.1. Post-Harvest Diseases and Physiological Disorders

There are many common diseases of pineapple such as bacterial heart rot (*Erwinia chrysanthemi*), root rot (*Pythium* spp.), fruit rot (*Erwinia carotovora*) and butt rot (*Chalara paradoxa*, *Ceratocystis paradoxa*, *Thielaviopsis paradoxa*). Plant viruses that commonly infect pineapples are bacilliform virus and closterovirus. Other diseases and physiological disorders of pineapple are summarized as in the Table 2 and Table 3, respectively.

3.2. Storage and Packaging

Post-harvest practices which are including the management and control of temperature and relative humidity of storage room, the selection and use of packaging and the application of post-harvest treatments such as fungicides and hot water treatment (Hossain and Bepary, 2015).

Table 2. Post-harvest diseases of pineapple with their associated plant pathogen and possible symptoms

Post-harvest diseases	Plant Pathogen	Symptoms
Fruit rot (black rot, water blister) (Rohrbach and Phillips, 1990).	*Thielaviopsis parado xa*, *Chalara paradoxa*, *Ceratocystis paradoxa*	• Stem will start to have black rot and gradually spread to the most of the flesh. • Flesh will show slight skin darkening as the initial external symptom. • The skin of rotted fruit flesh will be water-soaked and easily breaks as the flesh softens.
Yeasty fermentation (Lim, 1985; Rohrbach and Johnson, 2003)	*Saccharomyces* spp., *Hanseniaspora valbyensis*, *Candida intermedia* var. *alcoholophila*	• Yeast easily causes this fruit diseases through wounds of overripe fruit. • Fruit flesh turns bright yellow, softens and is cracked by large air cavities. • Severe inter-fruitlet corking symptoms with slight fruit cracking will be exhibited mostly on the green fruits.
Fruitlet core rot, leathery pocket and interfruit let corking (Sipes and Wang, 2017; Rohrbach and Schmitt, 1994)	*Penicillium funiculosum*, *Fusarium guttiforme*, *Candida guilliermondii*	• Fruitlet core will turn brown rot with having black spot and eye rot. • Fruits are misshapen and culled.
Pink disease (Rohrbach and Pfeiffer, 1976)	*Erwinia herbicola*, *Gluconobacter oxydans*, *Acetobacter aceti*, *Pantoea citrea.*	• Fruit has pinkish discoloration and wilted in appearance. • Internal flesh is easily water-soaked and light pink in color. • When canning, fruit tissue will show brownish-pink pigmentation when heated.
Green fruit rot (Pegg et al., 1995).	*Phytophthora cinnamomi*	• Fruitlet will develop water-soaked rot without showing any external symptoms during the initial stage of disease. • The green fruitlet with the initial water-soaked rot will progress to turn into an unique brown margin.
Marbling (Rohrbach and Schmitt, 1994).	*Acetobacter peroxydans*, *Acetobacter* sp., *Erwinia herbicola var. ananas*	• The color of the internal fruit tissue will develop from a yellowish to reddish brown and gradually changes to very dark or dull brown discoloration. • Infected fruit tissues harden with woody characteristics, become granular, breakable and freckled with color variants.

Table 3. Physiological disorders of pineapple with their associated causes and symptoms

Physiological disorders	Causes of physiological disorder	Symptoms
Endogenous Brown Spot (EBS) or Black Heart (BH) disorder (Abdullah, 1984; You-Lin et al., 1997)	Exposure of pineapples before or after harvest to chilling temperatures (< 7°C) for more than a week.	• Fruit flesh is water-soaked internally and symptoms are not easily noticeable externally. • Brown spots develop in the core area and gradually enlarge to the entire center portion.
Chilling injury (Antonio et al., 2004)	Exposure of pineapples to temperatures below 7°C.	• Fruit will fail to ripen properly and turn to dull green color when ripened. • Fruit core tissue will be darkened and susceptible to decay. • Crown leaves are easily wilted and discolored. • Broken internal tissue will cause watery appearance.
Sunscald or sun scorch (Keetch and Balldorf, 1979)	Effects of exposure of pineapples to strong sunlight or high temperature.	• Fruit turn whitish, scalded and bleached. • Fruit peel turns to yellow-white and gradually develops into pale brown with damaged flesh.
Cuts and bruises (Keetch, 1978)	Mishandling or rough handling of growers	• Damaged fruit flesh will develop into light straw-colored. • Fruit becomes translucent and cause leakage of cell contents. • Post-harvest losses caused by cuts and bruises at 60-100% and 40-60% respectively.
Flesh translucency/ porosity (Rohrbach and Paull, 1982; Paull and Reyes, 1996)	Greater fruit sensitivity to mechanical injury which begins before harvest and continues after harvest	• Fruits will have leakage of cellular fluids that can increase translucency that lead to increase pH, fruit weight and total esters. • This disease may develop into fully translucent green-shelled fruit.

Moreover, several factors such as a good selection of fruit at optimum stage of maturity, gentle handling of fruit, proper shade during fruit harvesting on the field, careful loading and packing of fruit in bulk bin or stackable containers and transporting of fruit by refrigerated truck are contributing to a promising resourceful post-harvest system.

3.2.1. Storage

A good storage of pineapples requires particular temperature, humidity or moisture and proper ventilation conditions. Air exchange rate at 40-60 times per hour with constant supply of fresh air is well recommended to remove gases arising upon the ripening and to reduce the carbon dioxide content of the hold air effectively (Medina and Garcia, 2005). Inadequate (danger of rotting) and excessive ventilations may be factors that cause spoilage of fruit due to growth of mould and bacteria during storage.

Most of the pineapple cultivars can be stored at 8-10°C for 4-5 weeks with a relative humidity of 90% (International TFNet, 2011). Besides, the quality of the pineapple can be maintained by using appropriate functional packing, transported in refrigerated container at proper control of temperature and equipped with filters for the control of ethylene gas until arrival to the final market (Medina and Garcia, 2005).

Several recommendations of refrigerated storage have been elaborated by Paull (1993) and Paull and Rohrbach (1985) with storage temperature from 7 to 12°C and 85 to 95% relative humidity (R.H.) for storage of pineapples fruit at the color break stage for 14 to 20 days and storage of ripe fruit at storage temperature from 7°C and 90% R.H. for 2 to 4 weeks, respectively. Generally, pineapples can be stored at 8 or 12°C for less than 3 weeks (Haruenkit and Thompson, 1996). Burden (1997) recommended pineapple to be stored at 7 to 13°C and 85 to 90% R.H. with 2 to 5% O_2 and 5 to 10% CO_2 for 2 to 4 weeks under controlled atmosphere storage. Furthermore, pineapples stored under hypobaric condition can help in extending the shelf life up to 30-40 days (Staby, 1976).

3.2.2. Packaging

Pineapples normally will be packed and contained in bamboo or rattan baskets, stackable plastic containers or corrugated fibreboard (CFB) boxes after harvested (International TFNet, 2011). Pineapples are cleaned properly by using blowers or brushed to remove the dirt, insects and any other foreign matters at the packing house. The peduncles are usually cut using a sharp knife to a length not exceeding 2 cm in order to prevent any bruises on other fruits caused by the protruded long peduncles. This is in accordance to the requirement of the Codex Standard (International TFNet, 2011). Then, the fruits are washed and submerged in disinfectant in trays. It is a common operation of submerging the fruits in solution (with Triadimefon), when the fruit is exported to United States and Europe (Medina and Garcia, 2005).

The cleaned pineapples are commonly packed in plastic boxes with holes in all sides for ventilation, in order to facilitate the heat flow produced from the fruits and exit the packaging material (Medina and Garcia, 2005). An attractive design of packaging material, presentation of the fruit and well-described content can definitely promote the fruit's sales and improve the income of the farmers. Pineapples are commonly packed in two methods, namely vertical packing and horizontal packing with the crown on the topside. As a way to prevent infection at the crown attachment point, pineapples are usually packed along with crowns to ensure better shelf life. Furthermore, the crown of each fruit is wrapped with perforated plastic sleeves to prevent water loss or yellowing of the crown.

The fruit packed in bamboo or wooden boxes is more prone to physical injuries due to sharp edge of bamboo and nails present in wooden boxes. To minimize the physiological weight loss, fruit should be packed in corrugated fiber boxes (Singh, 2009). A small numbers of uniform size and maturity of fruits around four or six packed are usually vertically or horizontally with a crown to fit closely the box, so that it can avoid physical compression and bruises on the packed fruits (Thompson, 2003). Each box can also be layered with sponge to protect the fruit base from further mechanical injury in the subsequent handling operations.

Depending on the market and carton size, fruits are normally packed to a net weight of 10 to 15 kg. In Malaysia, pineapples are packed using CFB boxes with net weight of 10 kg for exporting fruits to other countries including United Arab Emirates (UAE) and Saudi Arabia (International TFNet, 2011). For examples, large fruits may be packed up to 20 kg and high value small pineapples may be packed at 6 kg prior to shipping. The boxes used for the packing are cautiously stored in an insect-proof warehouse to prevent any pest or insect infestation on the fruits.

3.3. Post-Harvest Treatments

Apparently, there are several major challenges for pineapple industry in terms of prolonging storage life, lessening chilling injury, preserving quality and nutritional compounds in pineapple fruits. Therefore, many new and innovative techniques of preservation and post-harvest treatment are widely developed and explored. Common and latest developments in post-harvest treatments such as hot water treatment, fungicidal treatment, air-drying, waxing, pre-cooling, ultrasound and electrolyzed oxidizing water, UV-C light, salicylic acid treatment, calcium treatment and 1-methylcyclopropene (1-MCP) are elaborated accordingly.

3.3.1. Hot Water Treatment

Hot water treatment is commonly used to overcome the problem of pineapples from storage rot and being infected with mealybug, thrips and mites. A study done by Wijeratnam et al. (2005) on pineapples has shown that *Chalara paradoxa* inoculated pineapples at 10^4 spores/ml, followed by a hot water dip treatment at 54°C for 3 min were remained healthy when stored at 10°C for 21 days, followed by 48 h at an ambient temperature (28 ± 2°C) (Wijeratnam et al., 2005).

3.3.2. Fungicidal Treatment

Fungicides are often used to treat black rot of pineapples in many countries. Thiobendazole or Bavistin at 100 ppm is commonly applied on fruit as disinfecting agent to prevent growth of pathogens associated at pre-harvest stage. The length of fungicide dipping time is normally depending upon the size of the fruit and type of the fungicide used (Hossain and Bepary, 2015).

3.3.3. Air Drying

For the purpose of eliminating the excess water on the peel of the fruits, fruits normally will be subjected to air-drying process prior to wax or any other special treatments (Hossain and Bepary, 2015).

3.3.4. Waxing

Pineapples are commercially treated with a fungicide in a dip or spray application to control post-harvest fruit rot, caused by the fungus (Hossain and Bepary, 2015). A food grade wax, usually containing polyethylene/ paraffin or carnauba/paraffin- based, may also be applied to the fruit with the fungicide. The major advantage of waxing is to reduce the internal browning symptoms of chilling injury. It also reduces post-harvest water loss and improves fruit appearance (Paull and Lobo, 2012). The effectiveness of surface coating is greatly influenced by the type of coating materials and pineapple cultivars as reported by Zaulia et al. (2007). Waxing and surface coating prevent the access of oxygen to the internal tissue and may explain the effectiveness in reducing internal browning in stored pineapple.

3.3.5. Pre-Cooling

Pre-cooling is intended to lower the temperature of the freshly harvested produce instantly so that it can minimize the risk of spoilage due to shrinkage and microbial growth. Hence, it is one of the key operations to ensure the wholesome quality of fresh produce in the post-harvest system by keeping the salable produce fresh and desirable. Pre-cooling is advisable for prolonging the shelf life of pineapples due to its sensitivity to temperature. This operation is even a necessity for those fruits intended for export to distant market, which normally takes 2 to 3 days of harvesting. Thus, fruits are highly recommended to be cooled immediately, at least within 10 hours after harvesting time (Hossain and Bepary, 2015). Pre-cooling temperature for mature fruit is advised at 13-15°C for 6 to 8 hours. It varies and depends on the size of fruits and harvesting time. Forced air cooling system is more efficient compared to other pre-cooling methods, but this requires a specially designed unit incorporated with compatible packaging (Hossain and Bepary, 2015).

3.3.6. Ultrasound and Electrolyzed Oxidizing Water

The efficiency of the combined technologies using ultrasound (US) and electrolyzed oxidizing (EO) water on the shelf life and spore germination of *Fusarium* sp. isolated from pineapple cv. Phu Lae were investigated by Khayankarn et al. (2013). The spore suspensions were subjected to 1 MHz US irradiation and EO water at different frequencies of 108, 400 and 700 KHz for 0, 10, 30 and 60 min before incubated at 27°C for 2 days. Results presented that US irradiation at 1 MHz and treatments of EO water at all frequencies for 60 min gave the most effective suppression on spore germination of the fungus and extended the shelf life of pineapple up to 20 days.

3.3.7. Ultraviolet-C Light

There was a positive result gathered by Manzocco et al. (2016) using ultraviolet-C (UV-C) light treatments on the quality and shelf life of fresh-cut pineapple sticks. Radiance exposure of 200 J/m^2 UV-C light on the fresh-cut pineapple sticks packaged in conventional PET/EVOH/PE trays and stored at 6°C up for 15 days, exhibited higher reduction rate of lactic acid bacteria and yeast growths compared to the control. Therefore, there is a potential of using UV-C light treatment or incorporating this technology in the storage facility of fresh produce.

3.3.8. Salicylic Acid Treatment

A study was done by Lu et al. (2011) on the effect of pre-harvest salicylic acid spray or/and post-harvest salicylic acid immersion on the quality and internal browning of pineapple (*Ananas comosus* L. 'Comte de Paris') at 10°C for 20 days and at 20°C for two days. The authors found that the port-harvest internal browning incidence and intensity of the fruits have significantly decreased after subjected with all the salicylic acid treatments.

3.3.9. Calcium Treatment

Calcium can keep minimally processed fruits and vegetables longer in the fresh-like appearance by controlling the development of browning. Control of the flesh browning associated with chilling injury has been observed in the core and in the flesh adjacent to the core of Queen-type Mauritius pineapple (Hewajulige et al., 2003). Moreover, the incidence of the black heart disorder in Mauritius pineapple during field experiment was found to be reduced after fruits were subjected to two weeks intervals calcium chloride fruit spray in three split doses. A higher content of calcium was also detected in the core and flesh of pineapples treated with pre-harvest calcium fruit spray (Hewajulige et al., 2006).

3.3.10. 1-Methylcyclopropene (1-MCP)

The effect of 1-methylcyclopropene on black heart disorder in pineapple was studied by Selvarajah et al. (2001). Black heart disorder in pineapples stored at 10°C for four weeks were effectively controlled after treated with 1-methylcyclopropene at 0.1 ppm for 18 hours at 20°C.

4. Food Safety Concerns

There are adverse reactions documented associated with bromelain. Bromelain can induce IgE-mediated respiratory and gastrointestinal allergic reactions (Baur and Fruhmann, 1979). According to Gailhofer et al. (1988), bromelain is a strong sensitizer and causes sensitization due to inhalation and not to ingestion. Usually, bromelain allergy is occupationally acquired. Therefore, it is important to practise adequate precautions. In addition, bromelain may induce systemic reactions even at low concentrations. Pregnant women are strictly prohibited from taking the fresh pineapple juice to prevent premature induction of uterine contractions which may lead to abortions or miscarriages. However, consumption of pineapple juice during full term and onset of labor is encouraged to enhance oxytocin-induced uterine contractions as it is an uterotonic agent (Nwankudu et al., 2014). The other reported reactions include diarrhea, increased heart rate, nausea, vomiting, irritation of mucus membrane and menstrual problems. Besides, it is advisable to avoid overdosing in people with stomach ulcers, bleeding disorders, active bleeding or taking blood thinning medications. Apart from that, those with heart, liver or kidney diseases should take necessary precautions to avoid complications due to bromelain (Catherine, 2010).

CONCLUSION

The widespread availability and high nutritional value of pineapple have made it to be one of the most popular tropical and sub-tropical fruits. In addition, various studies have shown evidences that bromelain extracted from pineapple possesses therapeutic benefits to human health. Different parts of pineapple and its enzymes have also been exploited for industrial applications. Numerous post-harvest practices are developed and adopted to maintain the fruit quality, nutritional components and prevent losses. However, adequate precautions are necessary to prevent adverse reactions following consumption of pineapple, especially for pregnant women and those with stomach ulcers, bleeding disorders, heart, liver or kidney diseases.

REFERENCES

Abdullah, H. (1984). Black heart disease in pineapple: A review. *Food Technology in Malaysia*, *6*, 32-34.

Adikaram, N. & Abayasekara, C. (2012). In Pineapple.: D. Rees, G. Farrell & J. Orchard (Eds.), *Crop Post-Harvest: Science and Technology: Perishables*, (143-158). West Sussex, UK: Blackwell Publishing Ltd.

Adrover-López, P. A., Gonzalez, M. J., Miranda-Massari, J. R., Duconge, J. & Berdiel, M. J. (2016). Inflammatory sequelae after chikungunya virus infection: Proposed nutritional treatment. *Journal of Restorative Medicine*, *5*, 39-45.

Amini, A., Masoumi-Moghaddam, S., Ehteda, A. & Morris, D. L. (2014). Bromelain and N-acetylcysteine inhibit proliferation and survival of gastrointestinal cancer cells *in vitro*: Significance of combination therapy. *Journal of Experimental & Clinical Cancer Research*, *33*, 1-15.

Amini, A., Masoumi-Moghaddam, S., Ehteda, A., Liauw, W. & Morris, D. L. (2016). Potentiation of chemotherapeutics by Bromelain and N-acetylcysteine: Sequential and combination therapy of gastrointestinal cancer cells. *American journal of cancer research*, *6*, 350-369.

Amini, A., Masoumi-Moghaddam, S., Ehteda, A., Liauw, W., Akhter, J., Pilai, K. & Morris, D. L. (2015). Synergistic inhibition of human gastric and colorectal cancers by bromelain and N-acetylcysteine: An *in vivo* study. In: *Proceedings of the 106th Annual Meeting of the American Association for Cancer Research*, Philadelphia, PA, 18-22 Apr 2015; Cancer Research 2015; 75(15 Suppl): Abstract # LB-007.

Antonio, L. A. Jr. Takayoshi, A. & Tetsuya, T. (2004). Inhibition of chilling injury and quality changes in pineapple fruit with prestorage heat treatment. *Journal of Food, Agriculture and Environment*, *2*, 81-86.

Bartholomew, D. P., Paull, R. E. & Rohrbach, K. G. (Eds.), (2003). In: *The pineapple: botany, production and uses*. (1-301). Wallingford, UK: CABI Publishing.

Baur, H. & Fruhmann, G. (1979). Allergic reactions, including asthma, to the pineapple protease bromelain following occupational exposure. *Clinic Allergy*, *9*, 443-450.

Beattie, J. K. & Quoc, T. N. (2000). Manganese in pineapple juices. *Food Chemistry*, *68*, 37-39.

Brat, P., Hoang, L. N. T., Soler, A., Reynes, M. & Brillouet, J. M. (2004). Physicochemical characterization of a new pineapple hybrid (FLHORAN41 Cv.). *Journal of Agricultural and Food Chemistry*, *52*, 6170-6177.

Burden, J. N. (1997). Post-harvest handling for export. In: S. K. Mitra (Ed.), *Post-harvest Physiology and Storage of Tropical and Subtropical Fruits*, (1-20). Oxon, UK: CAB International.

Catherine, U. (2010). *Natural Standard Herb & Supplement Guide: An Evidence-Based Reference* (1st edition, 1-872). Missouri, United States of America: Elsevier/Mosby.

Chobotova, K., Vernallis, A. B. & Majid, F. A. A. (2010). Bromelain's activity and potential as an anti-cancer agent: current evidence and perspectives. *Cancer letters*, *290*, 148-156.

Chollakup, R., Tantatherdtam, R., Ujjin, S. & Sriroth, K. (2011). Pineapple leaf fiber reinforced thermoplastic composites: Effects of fiber length and fiber content on their characteristics. *Journal of Applied Polymer Science*, *119*, 1952-1960.

d'Eeckenbrugge, G. C. & Leal, F. (2003). Morphology, anatomy and taxonomy. In: D. P. Bartholomew, R. E. Paull & K. G. Rohrbach (Eds.), *The Pineapple: Botany, Production and Uses* (13-32). Wallingford, UK: CABI Publishing.

FAOSTAT. *World pineapple production*. (2017). Available from: http://www.fao.org/faostat/en/#data/QC.

Fitzhugh, D. J., Shan, S., Dewhirst, M. W. & Hale, L. P. (2008). Bromelain treatment decreases neutrophil migration to sites of inflammation. *Clinical immunology*, *128*, 66-74.

Gailhofer, G., Wilders-Truschnig, M., Smolle, J. & Ludvan, M. (1988). Asthma caused by bromelain: An occupational allergy. *Clinical & Experimental Allergy*, *18*, 445–450.

Gani, M. B. A., Nasiri, R., Almaki, J. H., Majid, F. A. A., Marvibaigi, M., Amini, N., Chermahini, S. H. & Mashudin, M. (2015). *In vitro* antiproliferative activity of fresh pineapple juices on ovarian and colon cancer cell lines. *International Journal of Peptide Research and Therapeutics*, *21*, 353-364.

Haruenkit, R. & Thompson, A. K. (1996). Effect of O2 and CO2 levels on internal browning and composition of pineapples Smooth Cayenne. In: *Proceedings of the International Conference on Tropical Fruits*, (343-350). Kuala Lumpur, Malaysia, 23–26 July 1996.

Hassan, A. & Othman, Z. (2011). Pineapple (*Ananas comosus* L. Merr.). In: E. M. Yahia (Ed.), *Post-harvest Biology and Technology of Tropical and Subtropical Fruits: Mangosteen to white sapote*, (194-212). Cornwall, UK: Woodhead Publishing Limited.

Hewajulige, I. G. N., Wilson-Wijeratnam, R. S., Wijesundera, R. L. C. & Abeysekere, M. (2003). Fruit calcium concentration and chilling injury during low temperature storage of pineapple. *Journal of the Science of Food and Agriculture*, *83*, 1451-1454.

Hewajulige, I. G. N., Wilson-Wijeratnam, R. S. & Wijesundera, R. L. C. (2006). Pre-harvest application of calcium to control black heart disorder in Mauritius pineapples during low-temperature storage. *Journal of the Science of Food and Agriculture*, *86*, 420–424.

Hossain, M. & Bepary, R. H. (2015). Post harvest handling of pineapples: A key role to minimize the post harvest loss. *International Journal of Recent Scientific Research Research*, *6*, 6069-6075.

International TFNet. *Pineapple: Post-harvest and Processing*. (2011). Available from http://www.itfnet.org/v1/2016/05/pineapple-post-harvest-processing/.

Joy, P. P. (2010). *Benefits and uses of pineapple. Pineapple research station*, Kerala Agricultural University. Kerala, India.

Keetch, D. P. (1978). *Bruising of pineapples. Farming in South Africa*, (2). Pretoria, South Africa: Bulletin H11/1978.

Keetch, D. P. & Balldorf, D. B. (1979). The incidence of certain pineapple fruit blemishes in the Eastern Cape and border. *Citrus Subtropical Fruit Journal*, *551*, 12–15.

Ketnawa, S. & Rawdkuen, S. (2011). Application of bromelain extract for muscle foods tenderization. *Food and Nutrition Sciences*, *2*, 393-401.

Khare, C. P. (2008). *Indian Medicinal Plants: An Illustrated Dictionary*. Berlin, Germany: Springer Science & Business Media.

Khayankarn, S., Uthaibutra, J., Setha, S. & Whangchai, K. (2013). Using electrolyzed oxidizing water combined with an ultrasonic wave on the post-harvest diseases control of pineapple fruit cv. 'Phu Lae'. *Crop Protection*, *54*, 43-47.

Lobo, M. G. & Paull, R. E. (2016). *Handbook of Pineapple Technology: Post-harvest Science, Processing and Nutrition.* Chichester, UK: John Wiley & Sons.

Lu, X., Sun, D., Li, Y., Shi, W. & Sun, G. (2011). Pre- and post-harvest salicylic acid treatments alleviate internal browning and maintain quality of winter pineapple fruit. *Scientia Horticulturae*, *130*, 97–101.

Majeed, M. & Borole, K. (2015). Evaluation of anti-inflammatory effect of pineapple juice in rheumatoid arthritis and osteoarthritis models in rats. *International Journal of Medical and Health Sciences*, *4*, 70-76.

Malézieux, E. & Bartholomew, D. P. (2003). In: D. P. Bartholomew, R. Paul & K.G. Rohrbach (Eds.), *The Pineapple: Botany, Production and Uses*. Wallingford, UK: CABI Publishing.

Manabo, C. A. & Frias, M. V. G. (2010). The antihelminthic efficacy of pineapple fruit mebendazole on soil transmitted helminthiases: a randomized controlled trial. *Pediatric Infectious Disease Society of the Philippines*, *11*, 35-43.

Manzocco, L., Plazzotta, S., Maifreni, M., Calligaris, S., Anese, M. & Nicoli, M. C. (2016). Impact of UV-C light on storage quality of fresh-cut pineapple in two different packages. *LWT - Food Science and Technology*, *65*, 1138-1143.

Maurer, H. (2001). Bromelain: biochemistry, pharmacology and medical use. *Cellular and Molecular Life Sciences*, *58*, 1234-1245.

Medina J. D. L. C. & Garcia, H. S. (2005). *Pineapple: Post-harvest Operations*, (25-28). Mexico: Food and Agriculture Organization of the United Nations.

Mohammed, M. (2004). *Optimizing post-harvest handling and maintaining quality of fresh pineapples (Ananas cosmosus (L.))*, (1-10). Port of Spain, Trinidad and Tobago: Morton Publishing.

Neto, A. R. S., Araujo, M. A., Souza, F. V., Mattoso, L. H. & Marconcini, J. M. (2013). Characterization and comparative evaluation of thermal, structural, chemical, mechanical and morphological properties of six pineapple leaf fiber varieties for use in composites. *Industrial Crops and Products*, *43*, 529-537.

Nwankudu, O. N., Ijioma, S. N. & Nwosu, C. O. (2014). *In vitro* investigation of fresh juices of ripe *Ananas comosus* (pineapple), *Carica papaya* (pawpaw) and *Citrullus vulgaris* (watermelon) for uterine contractile properties in non-pregnant rats. *International Journal of Zoology and Research*, *4*, 79-84.

O'Mahony, R. (2010). The antibacterial properties of dietary fruit. In: R. R. Watson & V. R. Preedy (Eds.), *Bioactive Foods in Promoting Health: Fruits and Vegetables*, (141-160). Amsterdam, Netherlands: Academic Press.

Paepatung, N., Nopharatana, A. & Songkasiri, W. (2009). Bio-methane potential of biological solid materials and agricultural wastes. *Asian Journal on Energy and Environment*, *10*, 19-27.

Paull, R. E. & Lobo, M. G. (2012). Pineapple. In: M. Siddiq (Eds.), *Tropical and Subtropical Fruits: Post-harvest Physiology, Processing and Packaging*, (1st ed, 333-357). New Delhi, India: John Wiley & Sons.

Paull, R. E. & Reyes, M. E. Q. (1996). Pre-harvest weather conditions and pineapple fruit translucency. *Scientia Horticulturae*, *66*, 59-67.

Paull, R. E. & Rohrbach, K. G. (1985). Symptom development of chilling injury in pineapple fruit. *Journal of the American Society for Horticultural Science*, *110*, 100-105.

Paull, R. E. (1993). Pineapple and papaya. In: G. Seymour, J. Taylor & G. Tucker (Eds.), *Biochemistry of Fruit Ripening*, (291-323). London, UK: Chapman & Hall.

Pauzi, A. Z. M., Yeap, S. K., Abu, N., Lim, K. L., Omar, A. R., Aziz, S. A., Chow, A. L. T., Subramani, T., Tan, S. G. & Alitheen, N. B. (2016). Combination of cisplatin and bromelain exerts synergistic cytotoxic effects against breast cancer cell line MDA-MB-231 *in vitro*. *Chinese Medicine*, *11*, 1-11.

Pavan, R., Jain, S. & Kumar, A. (2012). Properties and therapeutic application of bromelain: a review. *Biotechnology Research International*, *2012*, 1-6.

Pegg, K. G., Wassman, R. C., Broadley, R. H., Kelly, D. S. & Bagshaw, J. S. (1995). In: L. Coates, T. Cooke & D. Persley (Eds.), *Post-harvest Diseases of Horticultural Produce*, vol. *2*, *Tropical Fruit*. Brisbane, Australia: Queensland Department of Primary Industries.

Pieris, N. M., Jansz, E. R. & Dharmadasa, H. M. (2014). Extraction of bromelain from pineapple waste. *IAMURE Multidisciplinary Research*, *6*, 53-70.

Praveen, N. C., Rajesh, A., Madan, M., Chaurasia, V. R., Hiremath, N. V. & Sharma, A. M. (2014). *In vitro* evaluation of antibacterial efficacy of pineapple extract (bromelain) on periodontal pathogens. *Journal of International Oral Health*, *6*, 96-98.

Prior, R. L. & Cao, G. (2000). Antioxidant phytochemicals in fruits and vegetables: Diet and health implications. *Horticulture Science*, *35*, 588–592.

Rohrbach, K. G. & Johnson, M. W. (2003). Pests, diseases and weeds. In: D. P. Bartholomew, R. E. Paull & K. G. Rohrbach (Eds.), *The Pineapple: Botany, Production and Uses*, (203-252). Wallingford, UK: CABI Publishing.

Rohrbach, K. G. & Paull, R. E. (1982). Incidence and severity of chilling induced browning of waxed 'Smooth Cayenne' pineapple. *Journal of the American Society for Horticultural Science*, *107*, 453–457.

Rohrbach, K. G. & Pfeiffer, J. B. (1976). The interaction of four bacteria causing pink disease of pineapple with several pineapple cultivars. *Phytopathology*, *66*, 396–399.

Rohrbach, K. G. & Phillips, D. J. (1990). Post-harvest diseases of pineapple. In: R. E. Paull (Ed.), *Symposium on Tropical Fruit in International Trade*, (503-508). Honolulu, Hawaii: International Society for Horticultural Science.

Rohrbach, K. G. & Schmitt, D. P. (1994). Pineapple. In: R. C. Ploetz, G. A. Zentmyer, W. T. Nishijima, K. G. Rohrbach & H. D. Ohr (Eds.),

Compendium of Tropical Fruit Diseases, (45-55). Saint Paul, Minnesota: American Phytopathological Society.

Rosenberg, L., Krieger, Y., Silberstein, E., Arnon, O., Sinelnikov, I. A., Bogdanov-Berezovsky, A. & Singer, A. J. (2012). Selectivity of a bromelain based enzymatic debridement agent: A porcine study. *Burns*, *38*, 1035-1040.

Saha, S. K. (2016). *Effect of corticosteroid medication of periodontal and implant related procedures*, Master Thesis, University of Colorado at Denver.

Saradhuldhat, P. & Paull, R. E. (2007). Pineapple organic acid metabolism and accumulation during fruit development. *Scientia Horticulturae*, *112*, 297–303.

Secor, E. R., Carson, W. F., Singh, A., Pensa, M., Guernsey, L. A., Schramm, C. M. & Thrall, R. S. (2008). Oral bromelain attenuates inflammation in an ovalbumin-induced murine model of asthma. *Evidence-Based Complementary and Alternative Medicine*, *5*, 61-69.

Selvarajah, S. Bauchot, A. D. & John, P. (2001). Internal browning in cold-stored pineapples is suppressed by a post-harvest application of 1-methylcyclopropene. *Post-harvest Biology and Technology*, *23*, 167-170.

Sharma, G., Sharma, O. C. & Thakur, B. S. (2009). Section 1, Tropical and Subtropical Fruits: Pineapple. *In: Systematics of Fruit Crop*, (133-140). New Delhi, India: New Delhi Publishing Agency.

Singh, I. S. (2009). *Post-harvest handling and processing of fruits and vegetables*, (96-166). New Delhi, India: Westville Publishing House.

Singh, T., More, V., Fatima, U., Karpe, T., Aleem, M. A. & Prameela, J. (2016). Effect of proteolytic enzyme bromelain on pain and swelling after removal of third molars. *Journal of International Society of Preventive and Community Dentistry*, *6*, 197-204.

Sipes, B. & Wang, K. H. (2017). Pests, diseases and weeds. In: M. G. Lobo & R. E. Paull (Eds.), *Handbook of Pineapple Technology: Production, Post-harvest Science, Processing and Nutrition*, (1st ed, 62-88). Chichester, UK: John Wiley & Sons.

Staby, G. L. (1976). Hypobaric storage: An overview. *Combined Proceedings of the International Plant Propagation Society*, *26*, 211–215.

Thompson, A. K. (2003). *Fruit and Vegetables: Harvesting, Handling, and Storage*, (306-308). West Sussex, UK: Blackwell Publishing Ltd.

USDA. Guava. *National Nutrient Database for Standard Reference*: Release 28. (2017). The National Agricultural Library. Available from: https://ndb.nal.usda.gov/ndb/foods/show/2340?n1=%7BQv%3D1%7D&fgcd=&man=&lfacet=&count=&max=50&sort=default&qlookup=raw+pineapple&offset=&format=Full&new=&measureby=&Qv=1&ds=&qt=&qp=&qa=&qn=&q=&ing=.

Wijeratnam, R. S. W., Hewajulige, I. G. N. & Abeyratne, N. (2005). Post-harvest hot water treatment for the control of *Thielaviopsis* black rot of pineapple. *Post-harvest Biology and Technology*, *36*, 323–327.

You-Lin, T., Yu-Chan, Z. & Xing-Jie, T. (1997). A study on factors inducing and controlling post-harvest blackheart in pineapples. *Acta Horticulturae*, *425*, 595-603.

Zaulia, O., Suhaila, M., Azizah, O. & Mohamed Selamat, M. (2007). Effect of various coatings on the chemical changes of different pineapple cultivars (N36 and Gandul) at low temperature storage. *Journal of Tropical Agriculture and Food Science*, *35*, 107-120.

In: Tropical Fruits
Editors: C. Stewart Bogsan et al.
ISBN: 978-1-53612-885-7

Chapter 3

BROMELAIN FROM PINEAPPLE: ITS STABILITY AND THERAPEUTIC POTENTIALS

Sromona Das and Debasish Bhattacharyya*
Division of Structural Biology and Bioinformatics,
CSIR-Indian Institute of Chemical Biology, Jadavpur, Kolkata, India

ABSTRACT

Nature has provided strong proteolytic activity in the whole body of the pineapple *(Ananus comosus)* plant probably to strengthen its defense mechanism. The aqueous extract from the fruit, stem and leaf portions of pineapple are called 'fruit bromelain', 'stem bromelain' and 'leaf bromelain' respectively. They are rich in protease activity where the cysteine protease bearing the same name bromelain prefixed by the source dominates. Fruit and stem bromelain along with their isoforms have been extensively studied. Though similar but they are not identical. 'Bromelain' extracts also contain other cysteine proteases distinctly different from bromelain and an array of other enzymes like peroxidase, acid and alkaline phosphatase, amylase, collagenase, DNase, RNase, etc.

* Corresponding Author Email: dbhattacharyya1957@gmail.com; sromona0911@gmail.com.

'Bromelain' has been attributed with a number of therapeutic properties and it is believed that the enzymes present in it act in series to combat the diseases. Bromelain belongs to the class of kinetically stable proteins where a large energy of activation prevents the enzyme from being denatured under stressed conditions. It is also resistant to proteolysis. This ensures that those processes that require proteolytic activity of bromelain – irrespective of medical or industrial - can be continued over a period of time. Recent findings from our laboratory suggest that bromelain is a potent molecule that can destabilize protein aggregates and can inhibit formation of such aggregates from the monomeric and oligomeric state. Some of these proteins are involved in the development of amyloidogenic diseases. It appears that peptides derived from bromelain after protease digestion as par human digestive system are also effective in achieving this desired property. Thus, pineapple may be considered as a neutraceutical.

INTRODUCTION

Pineapple (*Ananas comosus*), the most edible member of the plant family Bromeliaceae is cultivated in tropical and subtropical countries. From an enzymologist's point of view, it is a rare cash crop with all parts of it's harvest having remarkably high proteolytic activity. In parallel, all portions of the plant find medicinal applications. Historically, presence of proteases in pineapple juice was first demonstrated in 1891 by Marcano [1, 2] and also by Chittenden [3] [later reviewed in 4, 5]. The crude aqueous extracts obtained from the stem, leaf and fruit of pineapple plant are termed as 'Bromelain' prefixed by the portion of the plant used in its preparation e.g., 'stem bromelain', 'leaf bromelain', 'fruit bromelain', etc. Distribution of high level of protease activity in the whole body of pineapple plant might act as a defense mechanism against soil microbes by digesting the protein components of the invaders. Of course alternate mechanisms to combat soil microbes by plants have also been evolved.

'Bromelain' is the collective name of a composite solution of different enzymes like thiol endopeptidases, phosphatases, glycosidases, peroxidases, cellulases, ribonucleases along with non-enzymatic components like protease inhibitor, glycoproteins, carbohydrates, organic acids, colored pigments and flavored compounds [4, 6-9]. The prevalence of all components except proteases being low to very low, 'bromelain' is widely known for its proteolytic activity. It is known for long that the concentration of 'bromelain' and thus the proteolytic activity increases as the plant matures and the protease activity is higher in stems compared to that of fruits [10]. This led to its commercial production and use as a phytochemical compound [4-10].

In general, 'bromelain' from all pineapple portions are constituted of a number of structurally and functionally closely related cysteine proteases. These proteases are also called bromelain prefixed by their sources like stem bromelain or fruit bromelain [11, 12]. Considering the closeness of the physico-chemical properties, these cysteine proteases may be termed as isoenzymes unless they show clearly distinguishable properties like non respondent to immunological cross-reactivity or distinct amino acid sequences. The description of bromelain enzymes is complex and confusing; sometimes they are referred in terms of molecular weight or isoelectric point or elution pattern from a chromatographic column [6]. Variation in composition of bromelain is dependent on the geographic location of the plant. Among all bromelain preparations, stem and fruit bromelain have been maximally studied. However, the medicinal property of pineapple is believed to be the synergistic effect of all the components present in 'bromelain' [13, 14].

Besides, the wide use of bromelain in food and beverage industries [15], cosmetic [16, 17] and textile industries [13], makes it a widely accepted therapeutic agent. The therapeutic potential of bromelain is on account of the biochemical and pharmacological properties of its constituent proteolytic enzymes, mainly a glycoprotein, which is present in addition to other insoluble minerals, coloured pigments, protease inhibitors, organic acids and solvents [18, 19]. The proteolytic enzymes in bromelain demonstrate *in vitro* and *in vivo*, fibrinolytic, antithrombotic [20], antiedematous [21], anti-tumor [22] and anti-inflammatory activities [23]. Enzymes of bromelain also regulate the activity of various immune cells and their cytokine production; modulate the functions of blood and molecules involved in endothelial cell adhesion [24] and show efficacy in anti-amyloidogenesis (discussed below).

COMPOSITION OF BROMELAIN

Stem and fruit bromelain - A detailed study on different proteolytic enzymes in crude stem and fruit bromelain was carried out in 1972 [25]. Two isoenzymes of bromelain were separated by isoelectric focusing resulting in a major basic component with an isoelectric point of 9.45 and a minor acidic component with an isoelectric point of 4.7 [26]. The basic glycoprotein with N-terminal Val or Ala and Mw of 28 kDa was designated as stem bromelain, while the acidic protein with Mw of 18 kDa and N-terminal amino acid Ala is designated as fruit bromelain. A number of reports suggested the presence of over six proteolytically active components in the stem (SBA and SBB 1-5) and two in fruit (FBA and FBB) depending on the geographic location of the plant [25, 27]. All active fractions analyzed by SDS-PAGE migrate as a single band with Mw of ~27 kDa for SBB 1-3 and ~23 kDa for SBB 4 and 5. A two-step cation exchange FPLC of crude stem bromelain revealed around eight basic proteolytically active enzymes with slight differences in Mws [28].

Stem bromelain (EC 3.4.22.32, formerly EC 3.4.22.33), a major enzyme present in the extract of pineapple stem is a cysteine protease of 23.8 kDa that belongs to the ($\alpha + \beta$) protein family [29]. It is the most abundant protein in pineapple. Its amino acid sequence shows high homology with papain [30], ervatamin [31, 32], chymopapain and actinidin [33] indicating presence of common folding patterns [11, 32]. X-ray diffraction studies of the available structures of bromelain reveal a high percentage of hydrophobic and uncharged amino acids with the polypeptide chain folding into two domains that interact through hydrogen bonds, salt bridges etc. [34].

Complete amino acid sequence of stem bromelain has been revealed [35]. It is a glycoprotein with one oligosaccharide moiety per molecule and exists as a single polypeptide chain with 211 or 212 residues, depending on the presence or absence of the N-terminal Ala. The total Mw of the glycosylated form of stem bromelain, with an oligosaccharide of 1 kDa, is around 23.8 kDa. The sequence of stem bromelain contains one sulfhydryl group (Cys_{26}) and three disulfide bonds between Cys_{23}-Cys_{63}, Cys_{57}-Cys_{96} and Cys_{152}-Cys_{199}. Bromelain though shares sequence homology with papain, its catalytic triad (Cys_{26}-His_{158}-Trp_{176}) has not been properly maintained as that of papain (Cys_{25}-His_{157}-Asn_{175}). More recently, two alternatives of catalytic 'triads', suggesting four catalytic residues for papain (Cys_{25}-His_{159}-Asp_{158}-Asn_{175}) and stem bromelain (Cys_{26}-His_{158}-Asp_{157}-Trp_{176}) have been proposed [36]. Assuming, essentiality of Cys_{26} and His_{158} residues and positions of disulphide bridges, minimum length of peptide chain that can accommodate the essential features of catalytic machineries is Cys_{23}-Cys_{199} (Figure 1A). This region corresponds to Mw of 18.99 kDa.

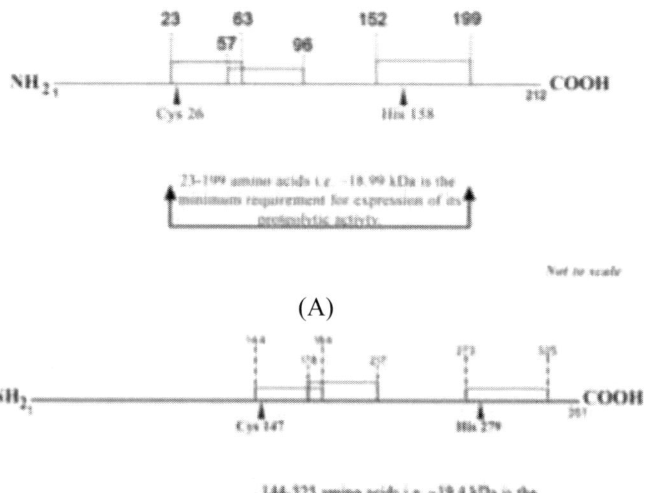

(A)

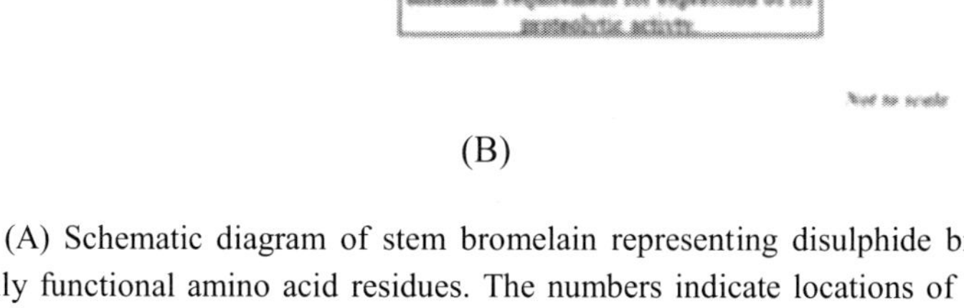

(B)

Figure 1. (A) Schematic diagram of stem bromelain representing disulphide bridges and catalytically functional amino acid residues. The numbers indicate locations of disulphide residues and the amino acid residues. (B) Same diagram as of (A) representing fruit bromelain.

Bioinformatics software, UNIPROT [37] provides the complete amino acid sequence of fruit bromelain (EC: 3.4.22.33), a glycoprotein with one oligosaccharide moiety per molecule and exists as a single polypeptide chain with 351 residues of 39.05 kDa. The sequence of fruit bromelain also contains one sulfhydryl group (Cys_{147}) and three disulfide bonds (Cys_{144}-Cys_{184}, Cys_{178}-Cys_{217} and Cys_{273}-Cys_{325}). Its catalytic triad comprises of Cys_{147} and His_{279}. Assuming, essentiality of these two residues and positions of disulphide bridges, minimum length of peptide chain that can accommodate the essential features of catalytic machineries is Cys_{144}-Cys_{325} (Figure 1B), corresponding to a Mw of 19.4 kDa.

Leaf bromelain and peel bromelain: Compared to fruit and stem bromelain, pineapple leaf, a common waste product of the plant received little attention in terms of its use in spite of its high proteolytic activity. This is more so for the pineapple crown leaf. Pineapple leaves are available in plenty but only at the site of cultivation. Marketed fruits carry crown leaves that account for the freshness of the fruits. Though studies reveal antimicrobial, thrombolytic and wound healing properties of the fruit and stem bromelain [24, 38-41], folk medicines also mention similar activities of crown leaf. Investigations on pineapple leaf extract in terms of different hydrolyzing enzymes and antimicrobial activity in order to determine its wound healing ability is pending.

STRUCTURE OF BROMELAIN

Acknowledging that the primary amino acid sequence of a protein ultimately leads to its tertiary structure and thus stability, the sequences of stem (212 residues) and fruit bromelain (351 residues) have been compared by Clustal W software. Stem and fruit bromelain are stabilized by three and four disulphide bridges respectively. When the sequences are aligned according to maximum homology, it appears that there are additional 121 and 34 amino acid residues at the amino- and carboxy- terminal ends respectively of fruit bromelain in comparison to stem bromelain (Figure 2A). It is not clear what this extra length of amino acids does in favor of fruit bromelain. Interestingly, when the extra sequences of fruit bromelain are deleted, the truncated fragment shows identical homology score as that of the intact sequences. However, sometimes in SDS-PAGE, the Mw of functional bromelain under certain experimental or preparative conditions appears below these required values as discussed above. It is likely that under those conditions the tertiary structure of bromelain remains grossly intact in spite of proteolysis of some of the hydrolysable bonds resulting in appearance of small functional fragments.

In the family of plant cysteine proteases, papain from Papaya latex serves as a model enzyme. Historically papain was one of the proteins whose x-ray crystallographic structure was solved in the early days of sixties [42]. Chymopapain, also abundant in papaya latex, is a distinct molecule from papain. Ficin from the latex of fig tree also belongs to this family. Ervatamin, from the plant *Tabernaemontana divaricata (Crepe jasmine/ Ervatamia coronaria)* is probably the most recent addition to this family and its crystal structure is known (PDB_ID_1IWD; http://www.rcsb.org/pdb/explore/explore.do?pdbId=1IWD) [43]. Stem bromelain shows maximum sequence homology with ervatamin B and not with papain (Figure 2B).

Compared to the volume of work done on bromelain on various aspects, it is rather unusual that its crystallographic structure is still unknown. Recently a structure of stem bromelain has been submitted in PDB (PDB_ID_1W0Q; http://www.uniprot.org/uniprot/P14518) [44] which is likely to be generated from bioinformatics software. Assuming that the sequence homology between stem bromelain and ervatamin B is appreciably high (61.16%), the structure of bromelain was generated using Swiss Model software (Figure 3). These two generated structures overlapped using PyMol software, indicates identical arrangenment among 85% of the folds of backbone structures while non-overlapping segments form either β-sheets or random coils (Figure 3).

(A)

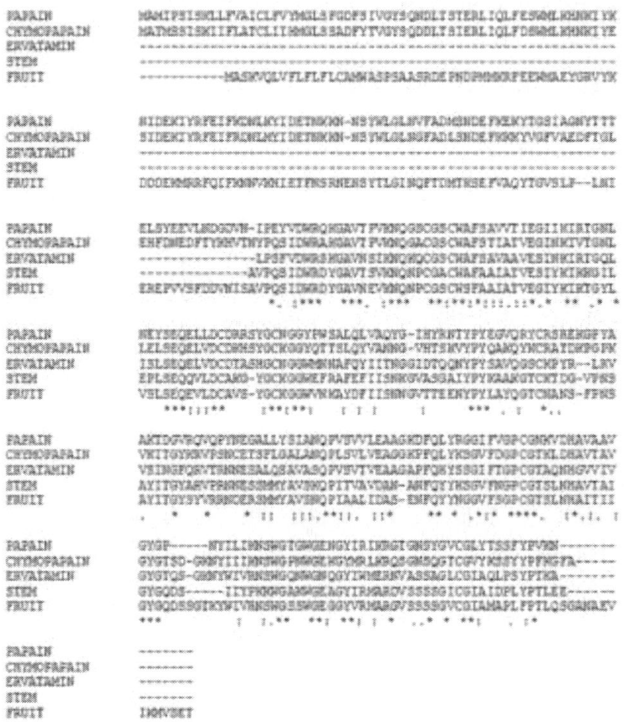

(B)

Figure 2. (A) Clustal W 2.1 sequence alignment between stem bromelain and fruit bromelain (Score: 68.3962). (B) Sequence alignment between papain, chymopapain, ervatamin B, stem bromelain and fruit bromelain. Out of all the pairing possibilities with that of stem bromelain and/or fruit bromelain, ervatamin reveals the maximum score of 61.1594 with stem bromelain. The symbols '*', ':', and '.' indicates identical, highly similar and similar residues respectively. No marking indicates dissimilar residues.

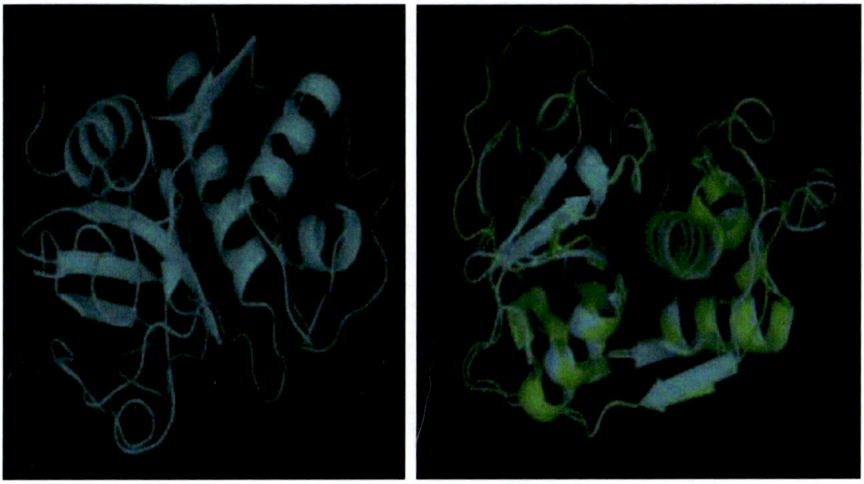

Figure 3. (A) In silico generation of the structure of stem bromelain using Swiss Model software and the crystal structure of ervatamin B (PDB registration ID: 1IWD) as reference. (B) Comparison of the generated stem bromelain structure as shown in (A) with the theoretical model of stem bromelain (PDB registration ID: 1W0Q).

PREPARATION OF BROMELAIN ON AN INDUSTRIAL SCALE

On account of the increasing demand of bromelain in industrial and medical markets, various purification techniques have been used to achieve maximum yield of the product at a reduced cost [13]. One should consider, especially in case of medical usage, that the preparation of bromelain must not exclude those components that synergistically act in curing the disease irrespective of the amount it occurs at source. This may not be important for an industrial preparation where proteolytic enzymes are major concerns. Ultrafiltration is a common method for extraction of bromelain from the juice of pineapple [8]. Commercially available bromelain is mostly extracted from the stem of pineapple through centrifugation, ultrafiltration, lyophilisation [45] and FPLC [46]. The purity of the product being an important factor, the crude mixture once extracted is subject to various purification methods to remove all impurities that reduce the specific activity of the enzyme and may also interfere in its mode of action

[47]. This definition of purification of bromelain may differ from an enzyme preparation. Several new advanced techniques that have been employed for the extraction and purification of bromelain are: different chromatographic techniques, membrane processes, precipitation, aqueous two-phase systems and reversed micellar systems. The advantage of stability of bromelain has been adapted in these processes (described below).

Chromatographic techniques - Several methods such as ion-exchange chromatography [48-51], high-speed counter-current chromatography [52], gel filtration chromatography [53], affinity chromatography and capillary electro-chromatography [54-56] are used for the process of purification. All these processes differ in their efficiency and yield [52], with ion-exchange chromatography being the most extensively employed technique due to its high specificity, low cost of matrix and reliability of the method in terms of protein recovery [57, 58]. Active recovery in the order of 87.4% is obtained when polyacryl acid bound iron oxide magnetic nanoparticles are used as a means to absorb bromelain from aqueous solution [54]. Immobilized metal affinity membrane results in a high recovery rate of 94.6% with around 15.4 fold purification factor of bromelain [59]. A 13-fold purification is achieved when bromelain is absorbed in expanded silver [60].

Purification through membrane filtration - This process utilizes membranes to purify molecules on the basis of size and shape. Utilizing principles of micro- and ultrafiltration, it is a quite useful method for enzyme separation and concentration [61]. The process yields an active recovery of 85% (through microfiltration) and an additional 10-fold concentration through ultrafiltration. As microfiltration and ultrafiltration followed by ammonium sulphate precipitation and ultracentrifugation are applied in a sequence, the yield increases to 98% [62]. Comparative studies of different processes together with membrane filtration show highest proteolytic activity of bromelain in that obtained from ultrafiltration [63].

Precipitation - Ammonium sulphate precipitation is an extensively used technique for protein purification [64]. A comparison between methods like ethanolic extraction, isoelectric focusing and ammonium sulfate precipitation reveal highest yield via ethanolic extract [65]. Also, recovery of bomelain from stem, bark and leaves of pineapple plant by ethanol, polyethylene glycol (PEG) and ammonium sulphate show maximum recovery through ethanol and minimum in case of PEG [66].

Aqueous two-phase system - This method consists of a reusable polymer [67] and a salt, or two polymers and is considered an important tool for the extraction and purification of biomolecules [68] due to its low cost, rapid extraction and ability to withstand high biomass load as compared to other separation techniques [69]. A high recovery rate of enzymes through this process is on account of the presence of PEG that causes alteration in the active sites of the enzyme. The method is widely used for the recovery of bromelain from the peel of pineapple [70].

Reverse micellar systems - While, micelles refer to aggregate of molecules that possess both polar and non-polar regions, reverse micelles are thermodynamically stable, minute surfactants that have an inner aqueous phase surrounded by an organic phase [71]. The method acts by entrapping the desired protein in the micelle while eliminating other impurities in the organic phase [72]. It has a high sample loading capacity, specific and easy to operate and is the ideal method for extraction even from diluted samples [73]. Studies on this method are still ongoing in order to increase the yield and purification fold.

Preparation of Fruit, Stem and Crown Leaf Bromelain on a Laboratory Scale

Laboratory processes of extracton involve usage of fresh pineapple cubes and stem. Fresh pineapple cubes are prepared by deleafing and peeling ripe fruits. Both stem and fruit should be crushed separately in suitable buffer in the ratio of 1:1 (w/v) to obtain the crude homogenized extract. A clear crude extract of both can be then obtained by centrifugation at 12857g for 10 mins at 4°C after precipitation of insoluble fibrous portions. The crude extracts thus obtained can be passed through a

Sephadex G-50 gel filtration column (fractionation range 5-30 kDa). Analysis of spectrophotometric results of size exclusion chromtaogaphy reveals two broad non-overlapping peaks, corresponding to Mws of 25.1 and 3.6 kDa respectively (Figure. 4). Proteolytic activity of the enzyme is restricted in both cases to the first fraction, while the second one correlates to non-proteolytic components like pigments. Analyses of various PAGE and SDS-PAGE electrograms reveal the presence of the protease in bromelain as a complex mixture of autodigested products and/or isoenzymes. While analysis of pooled proteolytic fractions show a single band in PAGE [74], reports of SDS-PAGE displayed two bands of 15 and 8.5 kDa formed on reduction of the 23 kDa molecule (Figure 4, A and B) [75]. To understand this difference in electrophoretic patterns, it is important to study the activity of bromelain based on gelatin (substrate) zymography, which affirms possible chances of noncovalent forces that hold these fractions together [74]. Concentrations of all pooled bromelain fractions can be studied by using $\varepsilon^{1\%}{}_{280nm} = 2.01$ [76].

Crowns of ripe pineapples excised from the fruit, leaves crushed for few minutes yield a homogenized extract. Similar to the preparation of fruit and stem bromelain, the extract can be centrifuged at 12857g for 10 min at 4°C in a suitable buffer in the ratio of 1:1 (w/v) until a clear supernatant is obtained [77].

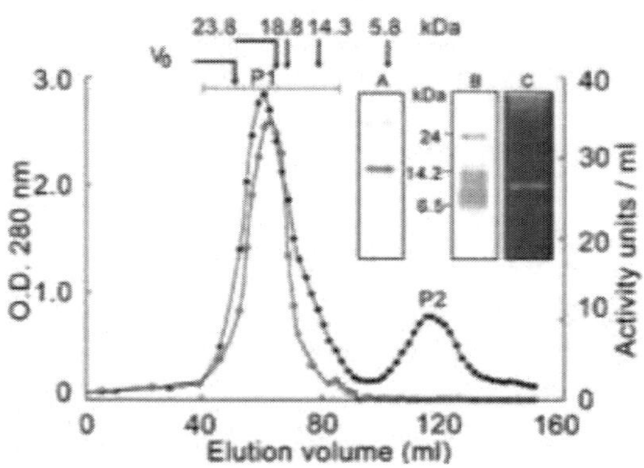

(A)

Figure 4. (Continued).

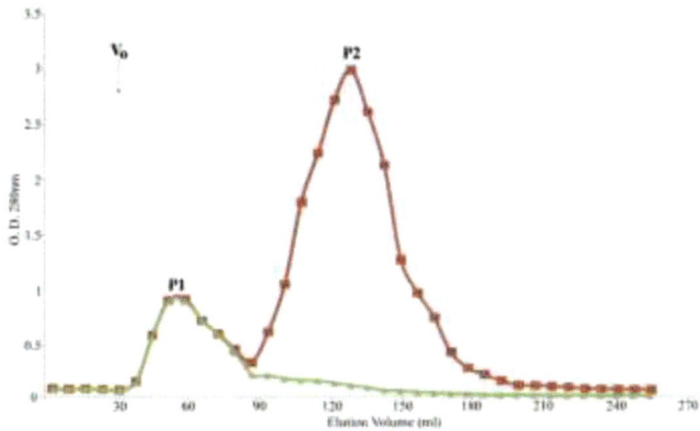

(B)

Figure 4.(A) Chromatographic separation of stem bromelain. The crude extract was lyophilized to a semisolid mass. The extract was taken in 2-ml of 10 mM Na-phosphate, pH 7.5 and centrifuged. It was applied to a Sephadex G-50 column (1.5 X 90 cm) and eluted with the same buffer. Elution was monitored at 280 nm (•). Proteolytic activity of the fractions was assayed using azocasein as substrate and was followed at 440 nm (◊). The void volume (Vo) and the positions of elution of the Mw markers used for calibration have been indicated at the top of the figure. Resolved fractions marked by the bars (P1 and P2) were collected, pooled and lyophilized. Inset: lane A, PAGE profile of the fraction P1; lane B, SDS-PAGE profile of P1 while the positions of the Mw markers have been mentioned at left; lane C, gelatin zymogram of P1. (B) Chromatographic separation of fruit bromelain. Experiment was done under identical conditions as of (A) except that 100 mg of crude extract was applied. It may be noted that the relative areas of P1 and P2 for stem and fruit bromelain are reversed. Abundance of protease in fruit bromelain varies depending on seasons, region and cultivars. (A) reproduced from [74].

NATURAL STABILITY OF BROMELAIN

All cellular proteins undergo continuous turnover. Protein degradation takes place via both lysosomal and non-lysosomal pathways. Different proteins display different half-lives ranging from few minutes to several days. Proteases, also being proteins, are susceptible to another pathway of degradation – autoproteolysis or autodigestion. A protease cannot retain its activity only on exogenous substrates without itself undergoing autolysis at the same time. In other words, a protease undergoes kinetic competition between selecting its natural substrate and its own counterpart that acts as a substrate undergoing autodigestion. Clearly in presence of plenty of substrate, the rate of autodigestion will be insignificant. Autodigestion also depends on the specificity of the protease and availability of susceptible bonds on the protease that acts as a substrate. Some proteases of the blood coagulation cascade possess very narrow specificity, for instance, factors VII, IX, X, XI, and XII, etc. are indefinitely stable against autodigestion due to the absence of susceptible peptide bonds that they can hydrolyze. Bromelain on the other hand, has a broad specificity of hydrolyzing nonpolar amino acid residues preferably at lysine, alanine, tyrosine, glycine, etc. Therefore, bromelain is expected to be affected by autodigestion unless its structure is protected from exposure of susceptible hydrolysable bonds.

Bromelain, being a protease of broad specificity, is prone to autodigestion. Concentration dependent studies on autodigestion of bromelain at pH 7.5 and 37°C show upto 2.12 times higher rate of autodigestion at 5 μg/ml (0.034 ± 0.005 min^{-1}) compared to that of 40 μg/ml (0.016 ± 0.004 min^{-1}) (Figure 5 and inset). In case of bromelain, autodegradation as observed from the estimation of residual protein content, increased with increase in protein concentration. Moreover, for 5 and 10 μg/ml, the autodigestion reaction attains a plateau at around 100 min, which is far earlier than for 20 and 40 μg/ml of bromelain that attains the same at ~250 min. A similar observation of concentration dependency was reported by Hale et al. [78]. However bromelain, from frozen fruit of *Bromelia balansae* Mez had no loss in activity when incubated at 37°C for a period of 120 min [79], and that of *Bromelia antiacantha* Bertol for 180 min [80].

The rate of thermal inactivation of stem and fruit bromelain follows a first order kinetics in the range of 55-60°C while it differs at higher temperatures of above 70°C. There still remains a controversy regarding the effect of pH and temperature on the stability of bromelain. Bromelain shows an increase of both activity and specific activity in the range of pH 5.7-7.0 and a subsequent decrease on further increase of pH. Temperatures ranging from 5 to 35°C at pH 7 show no significant effect [81]. While, bromelain is known to retain more than half of its proteolytic activity after incubation at 60°C for 30 min [81], loss in activity is noted when incubated for around 3 hrs or even during increase of temperature from 50 to 60°C [82].

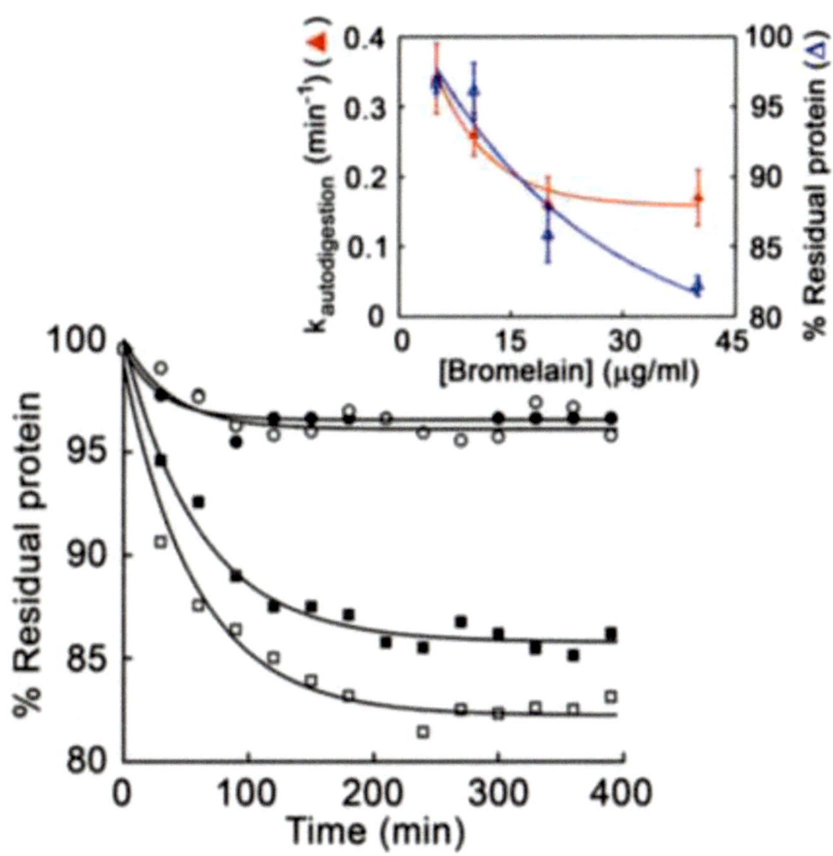

Figure 5. Autodigestion of bromelain at variable concentration as a function of time. Bromelain at 5 (●), 10 (○), 20 (■) and 40 (□) μg/ml were incubated at 37°C. At set time intervals, the residual protein was determined using Bradford reagent for protein estimation. Data points are the average of four closely spaced replicates. Inset: Rate of autodigestion of bromelain (▲) and the residual protein (Δ) content plotted as a function of its concentration. Reproduced from [74].

The stability of stem bromelain changes from the native state to a molten globule state at pH 0.8 via an acid-denatured configuration at pH 2.0 at 25°C. Though there is a loss of about 80% of native secondary structure and complete loss of tertiary structure at pH 2.0, bromelain exhibits native secondary structure with molten globular state and increased binding affinity to hydrophobic moiety like ANS at pH 0.8. Despite, the denatured structure depending on the change of the modifier states, the molten globule states differ significantly thereby making the latter a necessary intermediate for the denaturation of bromelain [13].

Preservation of pineapples following existing commercial protocols, leads to loss of proteolytic activity due to autodigestion and inactivation following harsh sterilization methods. Being an important edible fruit, it cannot be subject to chemical modifications for stabilization against autodigestion. However with higher concentrations (>50 mg/ml) rather than dilute solutions, partial stabilization can be achieved where surface denaturants dominate [81]. Investigations on whether supplementing large edible peptides, like casein, casein hydrolysate and aqueous extract of powdered pulses could stabilise bromelain indicated possible kinetic competition between hydrolysed extraneous peptides with that of autodigested ones.

Reports suggest hygienic preservation of cut fruits under frozen conditions as the best way to retain its activities. Proteolytic activity is retained in the order of 50% on storage of the fruits at 4°C for two months and it is improved to 75-80% when preserved at -4°C for upto 6 months without any pretreatment. A further lowering of storage temperature to -20^{0}C has no added advantage. Immuno-modulatory activity of fruit bromelain extract from fruit cubes stored at -4°C was comparable to fresh fruit extract. Since the fruit retains its physico-chemical properties during storage, its vitamin and mineral content remains unaffected. Though freeze burn is associated with some fruits as a mode of preservation, it shows no effect on stored pineapples with respect to color, texture, taste and flavor [83].

Microorganisms like that of coliforms, enterococci, lactis and yeasts as contaminants from soil, water, natural surface flora and natural handling dwell on fresh fruits [84]. Pineapple, possessing a pH range of 3.7 to 4.5 shows no microbial growth upto 24 hrs on incubation of its crude extract with inoculation of *E. coli* and/or yeast growth media, as an indication of antimicrobial properties of proteases [85-87].

Bromelain preparations are widely used in medicines, as laboratory reagents, and to a lesser extent in industry. In case of bromelain, its bioaffinity based oriented immobilization increases the thermal and pH stability compared to that of native bromelain [88]. To evaluate its efficacy as an industrial enzyme and to improve its stability by rational design, knowledge of its stability in terms of both the thermodynamic and kinetic aspects is required.

KINETIC STABILITY OF BROMELAIN

Stability of a Protein: Thermodynamic versus Kinetic

Thermodynamic stability alone does not guarantee that a protein will remain in the functional state during a given time scale since irreversible protein alterations may deplete the native state in a time-dependent manner. In an industrial setting, conditions are usually such that the functionality of a biocatalyst depends on the kinetics of irreversible loss of activity by some form of denaturation. A protein becomes irreversibly denatured if the partially or fully unfolded protein undergoes some permanent change such as aggregation or proteolytic degradation. In such cases, kinetic stability of the protein is determined by the activation energy of the unfolding process that precedes the irreversible step. Proteins that are stabilized with high potential energy of the transition state of unfolding are called ‘kinetically stable’ [89-91] (Figure 6). Enzymes of thermophilic organisms have acquired this property through the process of evolution.

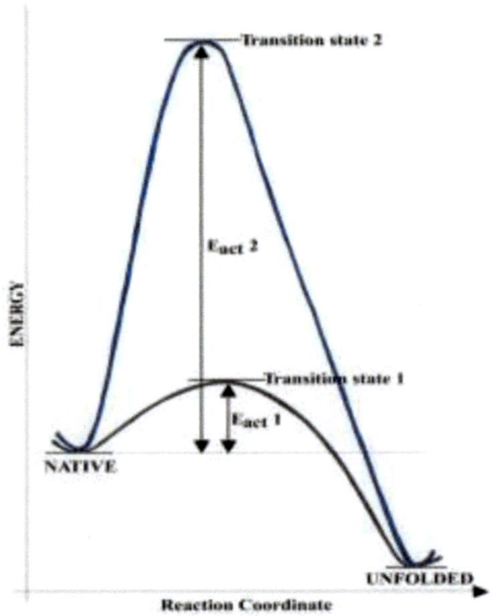

Figure 6. Schematic presentation of activation energy profile of mesophilic proteins (1) and kinetically stable proteins (2).

Kinetically stable proteins retain their native structure through exceedingly slow kinetics of unfolding rather than through a thermodynamic equilibrium favoring the native state (N) over the unfolded state (U). The presence of disulphide bonds, salt bridges and hydrophobic residues on the surface augment kinetic stability of a protein. Such proteins are less susceptible to proteolysis since the equilibrium between the N and U state seldom shifts towards the U state under native-like conditions, which undergoes further irreversible alterations leading finally to the irreversibly denatured protein. But, this depends on the relative amounts of N and U at equilibrium, which in turn depends on the rate of unfolding.

Proteins that survive under harsh external conditions e.g., in thermophilic (growth temperature 45-75°C) and hyperthermophilic (growth temperature ≥ 80°C) organisms or even plant enzymes surviving under desert conditions adapt this mechanism against denaturation [92]. These proteins exhibit remarkable thermal stability and resistance to chemical denaturants [93-95]. Though there is no consensus structural feature that could identify a protein to be kinetically stable, some generalization of their properties has been reported. In addition to high kinetic barrier to unfolding, these proteins are resistant to SDS binding and are stable against a wide variety of proteases. They are rich in β-sheet structure that allows bond formation among peptide stretches that are distantly situated along the length of the amino acid sequence. A dominant β-sheet structure partly explains high stability and overall compact rigid structure of the native state [91]. The kinetic stability of an enzyme leads to the persistence of activity even under denaturing conditions. In this context, encouraging is the long-term stability of fruit bromelain [53] and the structural rigidity of stem bromelain as demonstrated by its resistance to SDS binding [74] that questions whether the stability of bromelain originates from its slow rate of unfolding and satisfies the criteria of kinetically stable proteins (Table 1).

One of the primary requirements for a protein to be of industrial importance is its functional stability, which is governed by both thermodynamic and kinetic parameters. Naturally available enzymes are not always optimally suited for industrial applications accounting to the stability of the enzymes under processed conditions. Proteases were among the first enzymes to be targeted to obtain enzyme variants with improved stability, mainly because of the need for stable and functional variants in pharmaceutical preparations and application in detergents [96-100].

Table 1. Properties of stem and bromelain with respect to a kinetically stable protein

Features	Stem bromelain	Fruit bromelain
β-sheet content	17.1%	-
Disulphide bridge (Mw)	3 in 23.8kDa	3 in 39.05kDa
Energy of activation	29.3±0.03kcal/mole for inactivation70.66±0.54kcal/mol for unfolding	29.3±0.03kcal/mole for inactivation
Resistance to proteolysis	High	Lesser than that of stem bromelain
Resistance to SDS binding	Shows no change below 1mM	-

(-) indicates that no information is available.

Protein stability is often implicitly equated to the unfolding Gibbs energy at a given reference temperature, usually at 25°C. The stability of proteins that unfold completely and reversibly can be assessed in terms of ΔG of unfolding and are of great importance for understanding the principles of protein structure and stability [101]. Several denaturants act on the stability of acid unfolded and molten globule states as well as native bromelain [12, 76, 88, 102-108].

A study on the thermal denaturation of bromelain in acidic pH indicated that the process is consistent with an irreversible two-state model as demonstrated by differential scanning calorimetry experiments [103]. Immobilization offered more resistance to denaturation at higher temperatures of 60°C where the second phase of the two-phase process is prolonged by a factor of three [88]. Stem bromelain when exposed to increasing alkalinity exhibited conformational response through at least three different stages due to the ionization of tyrosine hydroxyl groups [76]. Equilibrium studies on acid induced denaturation of catalytically inactive stem bromelain suggested the existence of a partially folded intermediate state that retained about 42.2% of the native state secondary structure under low pH conditions [105]. This intermediate acquired different conformational states in presence of salt and alcohols and the alcohol-induced state showed a cooperative thermal transition in contrast to the non-cooperative thermal denaturation of the salt-induced state [106]. In comparison, catalytically active stem bromelain loses about 80% of its secondary structure and almost complete tertiary structure at pH 2.0 indicating that a modification of stem bromelain for inactivation affected the native to acid unfolded state transition.

These observations were better explained when presence of molten globulin state of bromelain was reported. Inactivation of the enzyme does not affect the stability of the molten globule state at pH 0.8 and the transition of the acid unfolded state to the molten globule state [12]. This molten globule state was stabilized only when subjected to a high concentration of low Mw polyethylene glycol [6]. A comparative study on the effects of pH, temperature and different organic solvents on the activities of glycosylated and deglycosylated forms of bromelain demonstrate that glycosylation contribute towards the functional stability of bromelain [104, 107]. This is a general feature that glycosylation of enzymes induces and thereby increases its stability. In fact, deglycosylated bromelain is more susceptible towards GdnHCl induced denaturation, indicating the importance of the carbohydrate moiety for stability of the enzyme [108]. During unfolding studies of bromelain by urea, guanidine and ethanol, four intermediate conformations; two with enhanced proteolytic activity, a third retaining native-like activity and a fourth non-functional intermediate possessing all the elements of secondary structure can be identified between the native and completely denatured state of bromelain [102].

Properties of Bromelain with Respect to Kinetically Stable Proteins

Stability against urea - Enzyme inactivation followed by unfolding, along with associated conformational changes provides a detailed view of the dynamics of the unfolding pathway. Transverse urea gradient zymography (TUGZ) makes use of a gradient of urea in combination with zymography. It is a useful means to study the processes of denaturation and inactivation of proteolytic enzymes simultaneously. It is an important method to analyze the folding dynamics of proteins, by following their denaturation and inactivation simultaneously [108-111]. It depicts a qualitative idea of the urea induced conformational changes that show a change in the mobility due to increase in the hydrodynamic radius of the unfolded pattern likewise changes in the net charge of exposed ionisable groups [112]. Rapid changes on the time scale of electrophoresis leading to equilibrium, expedites calculations of thermodynamic properties like ΔG, transition mid-point and cooperativity [113]. But, if the rate of interconversion is comparatively slow, then conformations can be kinetically resolved provided the electrophoretic separation is completed during the folding/unfolding transitions or in half of its time. Molecular modeling of stem bromelain, where the disulphide bonds are reduced, on a combined analysis with TUGZ pinpoints the evolutionary retention of Cys_{23}-Cys_{63} on account of the localized stabilization of this bond in the catalytic triad.

A fine line of equilibrium between the native and denatured states of a protein is based on the factors of electrostatics, hydrophobicity, temperature, etc. which support both structural and functional stability of the desired protein [114]. Reports on the study of TUGZ, support facts stating the stability of the active site containing domain and its resistance to urea-induced denaturation for stem bromelain [102]. It also provides an important means to investigate information relating to structural changes not captured by circular dichroism and solution state fluorescence measurements and acts as a two-dimensional system to assert denaturation and inactivation simultaneously. Studies show distinct visualization of the unfolding transitions of the different components present in the first proteolytically active fraction of stem bromelain [32]. One of the disadvantages of the method is that, on being inactivated upon unfolding the enzyme profile remains dark along with that of the substrate.

Thermal Stability

In case of a protease of broad specificity like bromelain, the molecule undergoes two phenomenons simultaneously at an elevated temperature – enhanced rate of autodigestion according to Arrhenius relation and thermal denaturation affecting tertiary structure of the molecule. When thermal denaturation precedes autodigestion, the protease remains intact but nonfunctional. On the other hand, if the protease remains functional at a higher temperature, it is almost certain that its structural integrity is maintained. At 45^0C, bromelain retained almost full activity at least for 1 hr indicating absence of autodigestion and thermal denaturation (Figure 7). However, at higher temperatures inactivation occur. Retention of native state like protein structure during the production of pineapple juice at 60°C possibly indicates the presence of nonfunctional enzyme [5]. Thus thermal inactivation preceded autodigestion.

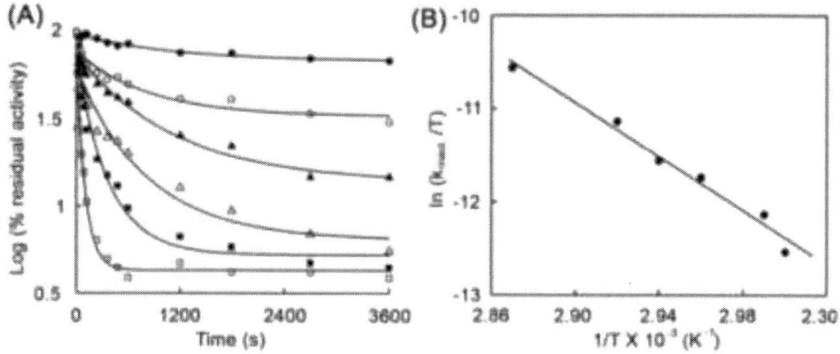

Figure 7. Temperature dependence inactivation bromelain. (A) Inactivation kinetics of bromelain at 45 (•), 50 (o), 55 (▲), 60 (Δ), 65 (■) and 75 (□) °C. Rate constants of inactivation of these reactions were determined after fitting the kinetic traces to a mono-exponential equation. (B) Eyring analysis of inactivation kinetics of bromelain (R^2 = 0.9787). Reproduced from [74].

Activation Barrier for Inactivation and Unfolding of Bromelain

The energy of activation (E_a) for inactivation (unfolding of catalytic site, E_a^{inact}) and unfolding of global structure (E_a^{unf}) of bromelain is determined from temperature dependent kinetics of inactivation and fluorescence emission change from aromatic amino acids exposed during the process of thermal unfolding respectively. The thermodynamic activation parameters, $\Delta S^‡$ and $\Delta H^‡$ as a measure of inactivation and unfolding free energy barrier are determined by monitoring the temperature dependence of its inactivation and unfolding kinetics respectively. Using the inactivation rate constants obtained from the kinetic traces at variable temperature (Figure 7 and 8), corresponding Eyring plot for inactivation can be constructed. The active site being a small part of the whole molecule, its unfolding can be easily studied by following the temperature dependent inactivation.

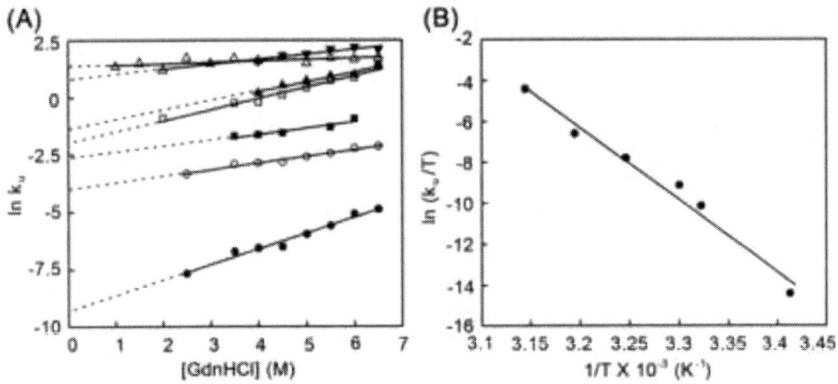

Figure 8. (A) Temperature dependent unfolding kinetics of bromelain in presence of variable concentration of GdnHCl. Kinetics of unfolding (K_U) was determined from the relative fluorescence obtained by dividing each data point by the first point of the native protein equilibrated at respective temperature. This ensured thermal denaturation of bromelain to be excluded at the temperature of experimentation before application of GdnHCl. Unfolding was followed 20 (●), 28 (○), 30 (■), 35 (□), 40 (▲), 45 (Δ) and 60(▼) °C. The unfolding rates under native like conditions are shown by extrapolating the dependencies to 0 M GdnHCl. (B) Eyring analysis of inactivation kinetics of bromelain (R^2 = 0.9777). Reproduced from [74].

The unfolding rates of the molecular structure of bromelain, however, cannot be affirmed using only temperature as a variant because of its structural rigidity. It is thus important to understand the temperature dependence of GdnHCl induced unfolding kinetics. The unfolding rate of a protein in the absence of a denaturant is determined by extrapolation of denaturant-dependent unfolding data. Analysis of results shows that both ln (k_u/T) and ln (k_{inact}/T) vary linearly with 1/T. Thermal inactivation is irreversible and follows a two-step mechanism revealing the presence of an intermediate state between the native and denatured states. The thermodynamic activation parameters of bromelain, $\Delta S^\ddagger$ and $\Delta H^\ddagger$ can be determined from the fit of the temperature-dependence of unfolding and inactivation rate constants in the absence of denaturant [74].The relations used are:

$$\ln(\frac{k}{T}) = -\frac{\Delta H^\ddagger}{R}\left(\frac{1}{T}\right) + ln\frac{k_B}{h}.\frac{\Delta S^\ddagger}{R} \qquad \text{(Eq. 1)}$$

Where, k_B is the, Boltzman's constant 3.03*10^{-23}cal/K, R the Gas constant and h is Planck's constant 1.58*10^{-34}cal-sec. A plot of $\ln(\frac{k}{T})$ versus $\frac{1}{T}$ produces a straight line with slope $-\frac{\Delta H^{\ddagger}}{R}$ and intercept$\frac{\Delta S^{\ddagger}}{R}$. Using the values for unfolding/inactivation, the respective values of activation energy can be estimated from the equation, $E_a = \Delta H^{\ddagger} + RT$.

Table 2. Thermodynamic activation parameters of stem bromelain at 25°C

Parameter	Unfolding	Inactivation
$t_{½}$ (weeks)	27.5	4.4
$\Delta H^{\ddagger}$ (kcal/mol)	70.07 ± 0.09	28.75 ± 0.14
$\Delta S^{\ddagger}$ (kcal/mol)	0.16 ± 0.01	0.01 ± 0.00
$\Delta G^{\ddagger}$ (kcal/mol)	22.39 ± 0.38	25.77 ± 0.53
E_a (kcal/mol)	70.66 ± 0.54	29.34 ± 0.23

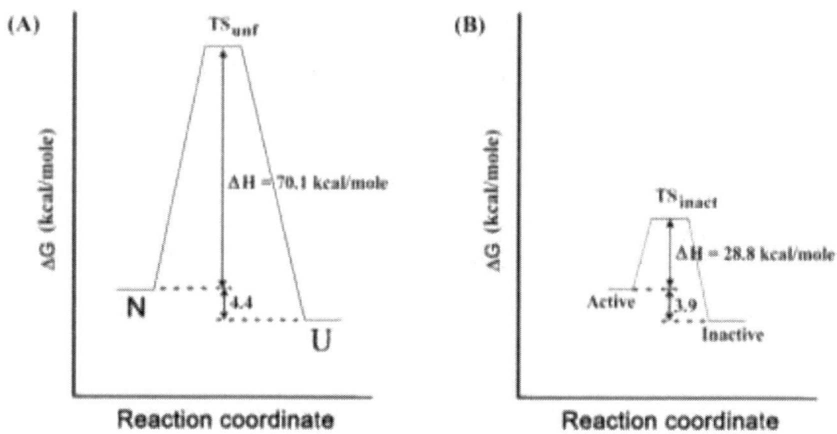

Figure 9. Energetics of (A) unfolding and (B) inactivation of stem bromelain. Reproduced from [74].

For bromelain to become susceptible to irreversible alterations, it must cross this activation barrier for unfolding. The native state k_u for bromelain has been found to be very slow with a half-life of ~27.5 weeks together with a very high value of E_a^{unf} i.e., 70.07±0.09 kcal/mole. Both E_a^{inact} and

E_a^{unf} of bromelain lie in the range of 24-120 kcal/mol, that kinetically stable proteins generally exhibit [115] (Table 2). Since inactivation generally precedes unfolding during protein denaturation [116], the fact that E_a for inactivation is less than E_a for unfolding is explainable. But the E_a^{inact} of bromelain also being in this range demonstrates the local stability of the active site. This indicates a high-energy barrier for unfolding and inactivation and hence the kinetic stability of bromelain. The energetics of unfolding and inactivation of bromelain have been summarized (Figure 9). It is to be noted that the effect of entropy on unfolding of stem bromelain is very low whereas it is almost insignificant in case of inactivation. Therefore, these processes are largely enthalpy driven.

Quantification of Protein Bound SDS by Rhodamine: A Measure of Kinetically Stable Proteins

Multicellular organisms generally cannot tolerate temperatures over 323 K (50°C). However, some organisms demonstrate optimal growth close to and even beyond this limit, for example, temperatures above 318 K (45°C) for thermophiles and above 353 K (80°C) for hyperthermophiles [117-119]. These proteins need to be particularly stable as a result of molecular adaptation to a particular physiological condition or to harsh environmental factors, such as high salinity, extreme pH and high temperatures. Among thermophiles, enhanced protein stability encompasses both thermodynamic and kinetic stability. While the kinetic stability depends on the energy barrier of unfolding, i.e., on the activation energy of unfolding (E_a), the thermodynamic stability is reflected in the conformational stabilities (ΔG_{unf}) which is up to 100 kJ/mol larger than those from mesophilic proteins [120], and in the midpoint transition temperatures for unfolding (i.e., T_m), which are typically between 20-30°C above those of mesophiles [121, 122]. Recent studies show that other proteins unrelated to thermophililc organisms also adapt this mechanism against denaturation [91]. These are also classified as 'kinetically stable proteins'. Because of their inherent stability, the native states of these proteins are by and large resistant to SDS binding. Conversely, resistance of a protein to bind SDS can be used as simple criteria to evaluate its stability. Quantifying the amount of SDS bound to proteins can assess this resistance.

Interaction of the fluorescent dye Rhodamine B (RhB) with free SDS in aqueous solutions is dependent on critical micelle concentration (cmc) of the detergent which rises by a ratio 2:1, once the cmc of the SDS solution is reached, which is around 7 mM in 10 mM Na-phosphate, pH 7.5 [123]. When RhB interacts with SDS bound to proteins, its fluorescence increases with increase in protein concentration [124]. Since an increase in protein concentration does not affect RhB fluorescence, this enhancement is solely dependent on the amount of bound SDS, which increases with increase in protein concentration. Four reference proteins, trypsin, ovalbumin, Bovine Serum Albumin (BSA) and Anti-diuretic Hormone (ADH) that are known to bind 1.4 mg SDS/mg of protein under fully unfolded conditions are generally included as control. Manning and Colon had demonstrated that papain and avidin are kinetically stable proteins that are resistant to SDS binding [91]. This indicates that amount of SDS bound to these proteins should be less than 1.4 mg SDS/mg protein at 25°C compared to when they are heated at 100°C in presence of SDS. Behavior of papain, bromelain and avidin with RhB was at par with their SDS resistance character under native conditions. This suggests inclusion of bromelain in the class of 'kinetically stable' proteins.

Resistance of Bromelain to SDS Binding

Surfactants are considered as non-specific denaturants of proteins like that of urea or GdnHCl that bind to proteins. Studies of protein-surfactant interactions are intriguing as they modulate the functional aspect of proteins that are largely detrimental and sometimes beneficial too [125]. On account of its use in industry, chemical, biological, pharmaceutical and cosmetic laboratories, study of such interactions for years reveal the nature of binding as monomers and/or as micelles depending on the nature of interaction and surfactant concentration [126-128]. Though proteins retain their structure and functionality in non-ionic detergents depending on their stability [129], in contrast, ionic surfactants bind in large amounts and denature membrane proteins and other soluble enzymes. SDS, used in gel electrophoresis for its denaturing effect, is one of the commonly used surfactant. Although several enzymes resist binding and unfolding ability of SDS, it is generally considered a denaturing surfactant.

Bromelain, belonging to the group of kinetically stable proteins that show resistance to SDS binding and proteinase K digestion, is rich in β-sheet structure and possess a rigid globular conformation [91]. It shows no interaction with SDS at low concentrations. Unlike most enzymes, bromelain has a very wide effective range of activity in both acidic and alkaline conditions, accounting for its activity in a variety of biological environments. The extent of bromelain's resistant to SDS binding has been systematically analyzed based on the secondary and tertiary structures, pyrene interactions, ITC isotherms and residual enzyme activity as a measure of the SDS concentration [53]. The cmc of SDS in presence of experimental conditions of bromelain is 5 mM. Significant masking/unmasking of aromatic residues accompanied by minor conformational changes, depicted by a decrease in fluorescence intensity and a shift of the $\lambda_{max,em}$ towards lower wavelengths can be observed beyond 1mM. While destabilization of electrostatic interactions in the native structure initiates the loss of tertiary structure, reconstructive denaturation measures the increase in secondary structure [130, 131].

Pineapple as Food

Bromelain when administered orally retains substantial proteolytic activity throughout the gastrointestinal tract. Although long term oral exposure to bromelain stimulates the development of high serum and stool anti-bromelain titers, the antibodies thus generated do not affect the proteolytic activity of bromelain within the gastrointestinal tract. These studies demonstrate that bromelain enzymes can be retained intact and in a proteolytically active form within the murine gastrointestinal tract as an indication of their stability [132]. On the contrary, administration of bromelain in humans results in higher total proteolytic activity, especially within the lumen of the gastrointestinal tract [133]. The proteolytically active bromelain though access the circulation, drop in plasma levels do not further affect the expression of cell surface molecules [134, 135]. However, bromelain concentrations within the gut lumen are clearly sufficient to remove at least a high percentage of bromelain-sensitive molecules present on cells exposed to the lumen [132]. This would include colon epithelial cells as well as lamina propria cells present in regions where surface integrity is compromised due to inflammation.

After harvesting, a ripe pineapple fruit can be preserved below 30°C in an airy place without stacking and avoiding sunlight for nearly 7 days. However, under identical conditions fruit cubes or juice are spoiled. Pineapples being a seasonal crop, huge quantities are harvested in a short time requiring their preservation. To our knowledge, processed fruit manufacturers sterilize the products under high temperature and pressure. A study reveals that the products are devoid of any enzymatic activity to the limit of detection with insignificant immuno-stimulatory activity. This is supported by the fact that SDS-PAGE analysis of the preserved fruit extract does not reveal any protein band but an abundant quantity of very small peptides. Preservation of the pineapple products under semi-frozen conditions at -4°C is a better cost effective option where all enzymatic activities are preserved for at least 30 days. An alternate proposition is to expose the products to γ-irradiation for sterilization that do not affect the biological activity. In short, the consumers who use canned or packed varieties of pineapple products should be aware that those items are of minimum medicinal values [83].

PHARMACOLOGY OF BROMELAIN

A number of remedial efficiencies of bromelain have been reported based on *in vitro* studies, *in vivo* observations on rodent models and also from clinical findings [20-24, 136]. These may be listed as follows:

- Reduces the amount of fibrinogen in blood, promotes fibrinolysis and activates plasmin
- Prevents blood platelet aggregation and adhesion to endothelial cells of blood vessels
- Acts as an anti-inflammatory agent thereby reducing the amount of prostaglandins E_2 and thromboxane A_2
- Increases secretion of various interleukins and of TNF-α
- Reduces and relieves osteoarthritis
- Promotes cytotoxicity of granulocytes against tumor cells
- Modifies pathways that support malignancy
- Increases tissue permeability for antibiotics
- Supports skin debridement during burns

Effect of Bromelain on Blood Coagulation and Fibrinolysis

Bromelain increases serum fibrinolytic activity in a dose-dependent manner by decreasing the synthesis of blood clotting protein, fibrin [137]. It is thus considered as an effective fibrinolytic agent that stimulates the transformation of plasminogen to plasmin resulting in increased fibrinolysis as an effect of fibrin degradation [4]. The decrease in proportion of serum fibrinogen in turn represses ADP induced platelet aggregation and retards the formation of prekallikrein and bradykinin at the site of inflammation [138]. This results in increased blood circulation at the site of injury, thereby reducing pain. Administration of bromelain in rats result in a dose-dependent reduction of fibrinogen, and at the highest concentration of bromelain it induces marked increase in thrombin and partial thromboplastin time [139].

Bromelain Prevents Platelet Aggregation

Attempts to isolate and characterize platelet aggregation inhibitory factors from bromelain show successful results [28, 140]. Platelet aggregation and cell adhesion properties of platelets to endothelial cells have been recently studied in more details [31]. Platelets incubated with bromelain prior to activation with thrombin, show a marked decrease in aggregation, while papain a similar protein is much less effective. Bromelain also reduces, *in vitro* the adhesion of thrombin-activated fluorescent-labelled platelets onto bovine aorta endothelial cells. Using an *in vivo* laser thrombosis model oral administration of bromelain to rats can significantly decrease the thrombus formation in mesenterial arterioles by 11% and in venules by 6%.

Effect of Bromelain on Immunogenicity

Bromelain is considered to be effective as a mode of adjuvant therapeutic approach in treatment of chronic inflammatory, malignant and autoimmune diseases [141]. Experimental evidences prove bromelain's efficacy in modulating surface adhesion molecules on T cells, macrophages and natural killer (NK) cells. It is also known to induce the secretion of Interleukin - 1β (IL-1β), IL-6 and tumour necrosis factor α (TNFα) by peripheral blood mononuclear cells (PBMCs) [142–148]. Besides, bromelain can block the Raf-1/extracellular-regulated-kinase-(ERK) - 2 pathway by inhibiting the T cell signal transduction [149]. *In vitro* experiments of bromelain define its ability in decreasing the activation of CD4 (+) T cells and reducing the expression of CD25 [150]. Moreover, there is also evidence that oral therapy with bromelain produces certain analgesic and anti-inflammatory effects in patients with rheumatoid arthritis, which is one of the most common autoimmune diseases [151].

Modifies Pathways that Support Malignancy

Recently evidences suggest that bromelain may be an effective anticancer therapeutic agent, as result of a systemic response involving multiple cellular and molecular targets, induced by this enzyme [152]. The molecular mechanisms underlying the anti-cancer activity of bromelain are still not fully understood. But certain studies have shown that bromelain has the capacity to modify certain key pathways that support malignancy. *In vitro* and *in vivo* studies governing anticancer activity of bromelain are mainly based on mouse and human cells, both cancerous and normal, treated with bromelain preparations. Beez et al. reveals that chemically induced mouse skin papillomas treated with bromelain decreases tumor formation, amount of preformed tumors and also causes apoptotic cell death [153]. While treatment of gastric carcinoma Kato III cell lines with bromelain significantly reduce cell growth [154], treatment of glioblastoma cells reduces its invasive capacity and de novo protein synthesis [155]. It increases the expression of apoptosis activators, p53 and Bax in mouse skin [153].

Signaling and overexpression of NF-κB play an important role in many types of cancers [156, 157]. Different studies have demonstrated the role of NF-κB, Cox-2, and PGE2 as promoters of cancer progression. Cox-2 is a multiple target gene of NF-κB that aids in conversion of arachidonic acid into PGE2 and accelerates the process of tumour angiogenesis and progression [158, 159]. Thus, obstructing the action of NF-κB, Cox-2 and PGE2 is a potential treatment of cancer. Bromelain is known to downregulate NF-κB and Cox-2 expression in mouse papillomas [153] and in models of skin tumourigenesis [160]. Bromelain also reduces the activity of cell survival regulators such as Akt and Erk, thereby promoting apoptotic cell death in tumours. It also inhibits bacterial endotoxin (LPS) induced NF-κB activity as well as the expression of PGE2 and Cox-2 in human monocytic leukemia and murine microglial cell lines [161, 162].

Reduces and Relieves Osteoarthritis

Osteoarthritis is a common form of disease in the present day. Women over the age of 45-50 years are mostly affected and the joints that are damaged from wear and tear are targeted first [163]. A combination of bromelain, trypsin and rutin is found to act similar to diclofenac in patients with osteoarthritis of the knee, both resulting in significant and similar reduction in the pain and inflammation [164]. Bromelain also acts as a harmless food supplement prescribed as an alternative treatment to nonsteroidal anti-inflammatory drug (NSAIDs) promising efficacy in rheumatoid arthritis and osteoarthritis [165]. It plays an important role in the pathogenesis of arthritis [166]. Bromelain has analgesic properties and displays direct effect on pain mediators such as bradykinin [167, 168]. Most recent reports state simultaneous use of bromelain with neutraceuticals such as turmeric in order to be effective in the treatment of degenerative joint pain diseases [169].

Promotes Cytotoxicity of Granulocytes against Tumor Cells

Studies of Bhui and co-workers [160] illustrating both long and short term investigations to analyse the basis of preventative (reduce in tumor volume by 65%) and anti-tumor-initiating potential of bromelain, pinpoints marked prevention against tumor initiation on carcinogen exposure, confirmed by delay in onset of tumorigenesis, and increased tumor free survival. The anti-tumor-initiating action of bromelain is hypothesized to be a result of the anti-inflammatory activities of bromelain. Dhandayuthapani et al. [22] confirmed the dose-dependent cytotoxic effects of bromelain in the GI-101A breast cancer cell line, on the basis of studies that mark the increased level of caspase-cleaved cytokeratin

(CK18) marker indicating cell death via apoptosis. Bromelain is thus suggested to increase apoptosis related dell death in breast cancer cells with increasing concentrations.

Bromelain Increases Tissue Permeability for Antibiotics

Antibiotic activation is one of the main uses of bromelain for several years. Bromelain can alter the permeability of tissues and organs to various drugs. It has been known to increase tissue permeability of penicillin and tetracyclins following oral administration. This facilitates absorption and also enhances diffusion after subcutaneous and intramuscular application of the antibiotics, with reduced side effects and higher serum and tissue levels [170, 171]. Studies on human, show bromelain's ability to increase blood and urine levels of antibiotics and also results in higher blood and tissue levels of amoxicillin and tetracycline when administered simultaneously with bromelain [172]. Even half a century before a compiled study showed that patients affected with pneumonia, bronchitis, staphylococcus infections, thrombophlebitis, pyelonephritis and rectal abscesses, healed faster on treatment with a combination of antibiotics and bromelain than that with antibiotics alone [173]. Similarly rhinosinusitis, inflammation of nasal cavity and paranasal sinuses [174] and chronic rhinosinusitis (CRS), disruption of mucus membranes, is treated more effectively with bromelain, a mucolytic agent (83%) versus the placebo group (52%) [175]. Bromelain decreases the formation of pro-inflammatory prostaglandin and reduces swelling in nasal passages [176] and aids its drainage [177].

Bromelain Supports Skin Debridement during Burns

Eschar, the dry skin formed on wounds, especially after burns enhance chances of infection [178]. Bromelain contains non-proteolytic enzyme, escharase, beneficial for the debridement of the necrotic scar tissue, accelerated wound healing [24, 179] and quick reepithelialisation [179-182]. Researchers have found bromelain to be useful in healing post-surgical wounds and lessening their ache and swelling [21, 183]. In two different studies carried out in porcine model, one using application of bromelain gel (35% bromelain in a lipid base) and another using topical bromelain, the former shows rapid removal of the necrotic tissue and preservation of unburnt regions [178, 182, 184], while the latter promotes the recovery of blood perfusion, partial pressure of oxygen in tissues and controls the expression of TNF-α [185].

BROMELAIN AS A PROSPECTIVE ANTI-AMYLOIDOGENIC AGENT

Formations of fibrillar structure of proteins with serious medical consequences are prominently emerging. Many diseases ranging from neuronal disorder in brain to cataract formation in eyes to rheumatoid arthritis are related to amyloid deposition of specific proteins or peptides. One of the characters of these hard, insoluble masses (plaques) is their resistance to hydrolysis by available proteases *in vivo* leading to difficulties of their clearance. Since bromelain is of broad specificity and a foreign protease to mammalian body, we investigated whether it can degrade fibrillar structures of proteins. For a preliminary estimation, the fluorophore thioflavin T can be employed to estimate retention of preformed amyloid-like structures of BSA in presence of bromelain. What is surprising is that, not just intact functional bromelain but also its inactive form and even its autodigested products devoid of the original intact molecule can induce destabilization of pre-aggregated structure of BSA (Figure 10). The pool of peptides generated from bromelain after proteolysis by human digestive enzymes like pepsin, trypsin, chymotrypsin, elastase, carboxypeptidases and animo peptidases under appropriate conditions, also offered similar potency. This clearly indicate that the process of destabilization is not mediated through proteolysis rather by one or more stretch of amino acid sequences either present on the surface of bromelain or generated after autodigestion are the causative agents (Figure 11). This assumption appears to be correct as the process of destabilization of BSA aggregate is dependent on the concentration of the bromelain peptides (Figure 12A). Since ThT assay or SE-HPLC profiles are indirect quantification of fibrillar structures, different molecular states of BSA can be viewed under transmission electron microscopy. While the monomeric state show presence of a few small oligomers only, the aggregated state possesses a dense mesh like network. When this entity is treated with digested bromelain peptides, the network disappears forming small fragments. Visible structures include a few fragments of the network, some oligomeric states of variable size, very small multimers and monomer sized molecules (Figure 12 B-D). Prolonged incubation completely removes the larger fragments.

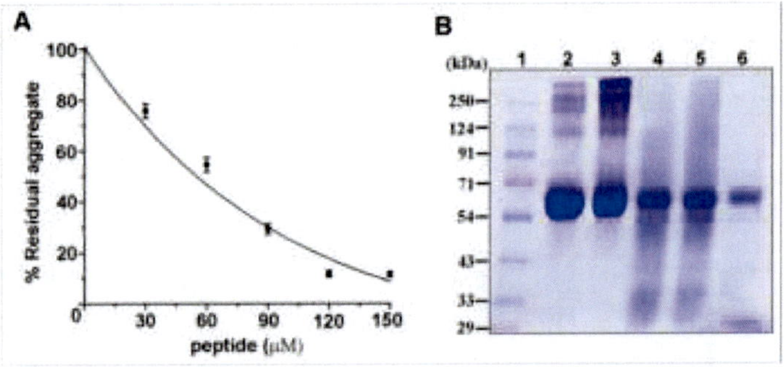

Figure 10. (A) Destabilization of preformed BSA aggregates in presence of variable concentration ofbromelain derived peptide pool. This pool of peptides was generated from autodigestion of stem bromelain. Disaggregation was monitored after 24 hrs of incubation in presence of the peptides at 37°C using Thioflavin-T fluorescence assay (ex: 450 nm; em: 482 nm). In absence of the peptides, the aggregate remained stable under the experimental conditions for at least 24 hrs and the corresponding emission intensity was considered as 100%. (B) SDS-PAGE profile of destabilization of BSA aggregate in the presence of bromelain peptide pool. Lane-1, Marker proteins whose Mws have been indicated, Lane-2, BSA monomer (1 mg/ml) which is contaminated with dimmers and above, Lane-3, BSA aggregate (1 mg/ml) where the abundance of the multimers has been increased; Lane 4-6, BSA aggregate after incubation with 20-120 μl of bromelain where the multimers have been completely disappeared. It may be noted that BSA aggregate that are linked by noncovalent forces are likely to be dissociated by SDS. Reproduced from [136].

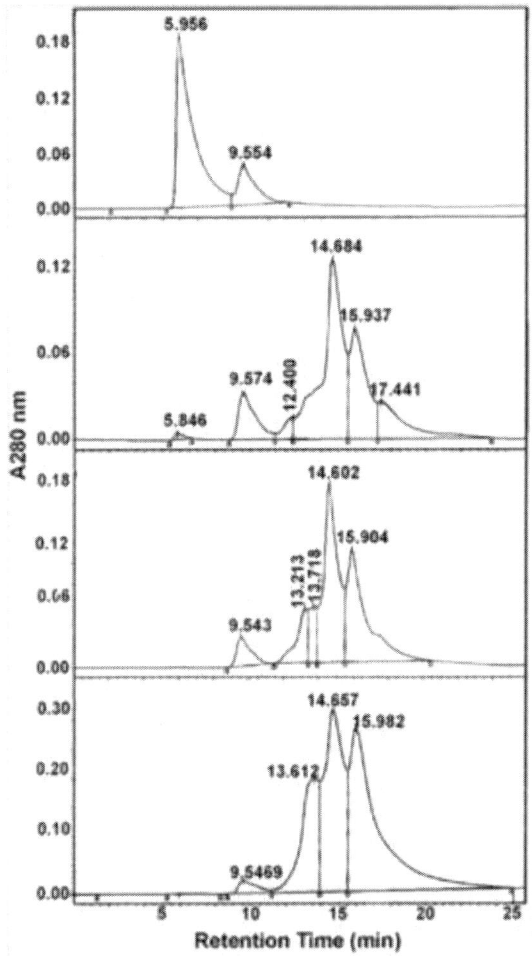

Figure 11. Degradation of BSA aggregates by bromelain derived peptide pool as followed by SE-HPLC analysis. BSA aggregate was incubated with 0, 30, 60 and 120 μM of peptides for 24 hrs and at 37°C and was applied to Waters Protein Pak 125 SE-HPLC column equilibrated with 20 mM Na-phosphate, pH 7.5 containing 0.2 M NaCl (upper to lower panel). Flow rate was 0.8 ml/min and elution of the components was followed at 280 nm. The component corresponding to Rt = 15.937 is the peptide pool. The aggregate of BSA and its monineric state appeared at Rt = 5.956 ± 0.03 min and 14.684±0.02 respectively. Components eluting at intermediate retention times were oligomers of variable multimericities. The profiles indicate disappearance of the aggregate and higher oligomers with increase of peptide concentration. Reproduced from [136].

Though the mechanism/s of these interactions remain uncertain at this moment, it is clear that the peptides induce irreversible change of configuration of BSA that make them unable to form an aggregate once dissociated from the fibrillar structure. Assuming that the general features of protein aggregation are similar, there remains a fair possibility that bromelain derived peptides may also destabilize aggregate prone proteins that are of clinical importance. Investigations in that direction is quiet impressive [136, 186, 187]. It is noteworthy that peptides derived from bromelain can also inhibit formation of amyloid-like fibrillar structure of proteins from their nomoneric state under conditions when protein aggregates are formed. Thus both prevention and cure of certain diseases are suggestive.

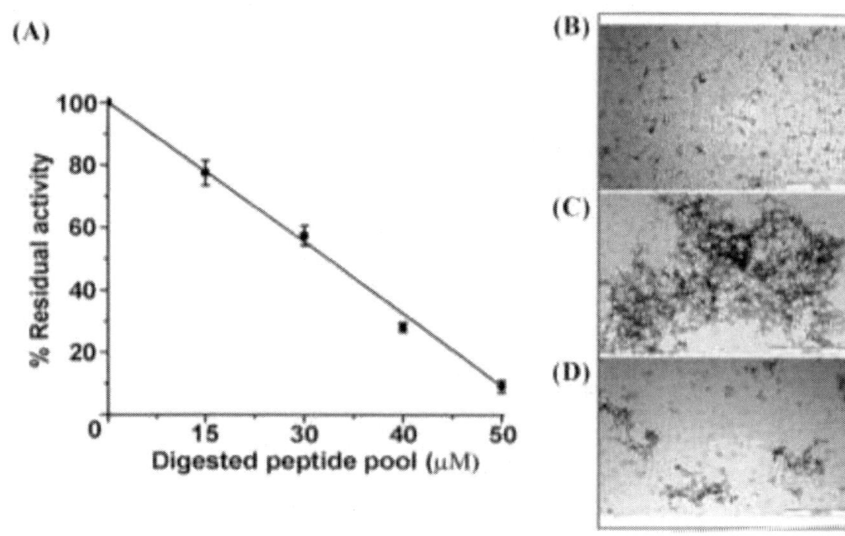

Figure 12. (A) Destabilization of preformed BSA aggregates in presence of variable concentration of bromelain derived peptide pool. This pool of peptides was generated from fruit bromelain as par hydrolytic pattern of human digestive system (described in the text). Residual aggregate structure was estimated from Thioflavin T assay. Transmission electron micrographs of different molecular states of BSA: (B) monomeric form; (C) aggregates in 10 mM Na- phosphate, pH 7.5 containing 100 mM NaCl and (D) after incubation of preformed aggregates with 50 μM of digested peptide pool at 37°C for 24 hrs. In each set, 10 μl aliquots of samples were taken for negative staining. Images are shown relative to calibration bars of 1000 nm. Reproduced from [136].

It is now fairly well established that bromelain or peptides derived from it can irreversibly disaggregate preformed human insulin aggregates and can inhibit formation of such aggregates from the monomeric state of insulin. Cellular studies with synthetic des-octapeptide of insulin that promotes aggregation but is devoid of hormonal activity show similar behaviour in presence of bromelain or peptides derived from it [187].

TOXICITY OF BROMELAIN

In acute and subacute toxicity tests using rat model, no signs of toxicity or mortality can be observed when bromelain from crown leaves is administered, indicating that there is a wide margin of safety for oral administration of bromelain [188]. Based on this observation, an aqueous extract from the crown leaves of pineapple, which contains fruit bromelain as the major protein constituent, is usually recommended in the treatment of microbial infection and also as a wound healing agent. Bromelain has very low toxicity with an LD_{50} (lethal doses) greater than 10 g/kg in mice, rats and rabbits [189]. Dosages of 1500 mg/kg per day when administered to rats neither provoke any alteration in food intake, histological and hematological changes parameters nor induce any carcinogenic and teratogenic effects [170]. After application of bromelain (3000 FIP unit/day) to human over a period of ten days, no significant changes in blood coagulation parameters can bee observed [146].

CONCLUSION

Traditional medical use of the extract derived from different parts of the pineapple plant called ‘bromelain’ is known across the world. Medicinal properties of ‘bromelain’ are attributed primarily to the proteases and secondarily to other biomolecules of lower abundance present in it. With no detectable negative impact on health after prolonged use, it is used as an effective supplement for health. Bromelain, the cysteine protease from pineapple bearing the same name has now been categorized as kinetically stable where a large energy of activation for denaturation protects the molecule from being inactivated under relatively harsh conditions. Stability and broad substrate specificity of bromelain indicate its potency as an industrial protease. Huge amount of inedible part of pineapple plant like stem, leaf, peel and crown are treated as agricultural waste. With the advancement of separation techniques, advantage of stability of the enzymes in 'bromelain' and urge to create a green earth, there remains immense potency to recover these enzymes for therapeutic and industrial purposes.

ACKNOWLEDGMENTS

SD was supported by a fellowship from CSIR network project (mIND, BSC0115). We are indebted to Dr. Reema Bhattacharya, Dr. Debratrna Mukherjee, Dr. Sangita Dutta and Dr. Kanika Sharma for initiating bromelain research in our laboratory.

REFERENCES

[1] Marcano, V; Phar, BU. *PDVSA article citing Vicente Marcano* (in Spanish), 5, 77 (1891).

[2] Lotz-Winter, H. On the pharmacology of bromelain: an update with special regard to animal studies on dose-dependent effects. *Planta Med.*, 1990, 56, 249-253.

[3] Chittenden, RH; Elliott, PJ; Meara, FS. On the ferments contained in the juice of the pineapple (*Ananassa sativa*): together with some observations on the composition and proteolytic action of the juice. *Transac Connecticut Acad Arts Sci.*, 1892, 8, 281–308.

[4] Balls, AK; Thompson, RR; Kies, MW. Bromelin: properties and commercial production. *Ind Eng Chem.*, 1941,33, 950–953.

[5] Taussig, SJ; Batkin, S. Bromelain, the enzyme complex of pineapple (*Ananas comosus*) and its clinical application: an update. *J Ethnopharmacol.*, 1988,22, 191–203.

[6] Nadzirah, K; Zainal, S. Efficacy of selected purification techniques for bromelain. *Food Res J.*, 2013, 20, 43–46.

[7] Ahmad, B; Ansari, MA; Sen, P; Khan, RH. Low versus high molecular weight poly(ethylene glycol)-induced states of stem bromelain at low pH: stabilization of molten globule and unfolded states. *Biopolymers*, 2006, 81, 350–359.

[8] Pavan, R; Jain, S; Shraddha, KA. Properties and therapeutic application of bromelain: a review. *Biotechnol Res Int.*, 2012, 2012, 976203.

[9] Hennrich, N; Klockow, M; Lang, H; Berndt, W. Isolation and properties of bromelin protease. *FEBS Lett.*, 1969, 2, 278–280.

[10] Heinicke, RM; Gortner, WA. Stem bromelain—A new protease preparation from pineapple plants. *Economic Botany*, 1957, 11, 225–234.

[11] Rowan, AD; Buttle, DJ. Pineapple cysteine endopeptidases. *Methods Enzymol.*, 1994,244, 555–568.

[12] de Lencastre Novaes, LC; Jozala, AF; Lopes, AM; de Carvalho Santos-Ebinuma, V; Mazzola, PG; Pessoa Junior, A. Stability, purification, and applications of bromelain: A review. *Biotechnol Progress.*, 2016, 32, 5–13.

[13] Arshad, ZIM; Amid, A; Yusof, F; Jaswir, I; Ahmad, K; Loke, SP. Bromelain: an overview of industrial application and purification strategies. *Appl Microbiol Biot.*, 2014, 98, 7283-7297.

[14] Anwar, T; Ahmad, B; Younus, H. Cross-linked stem bromelain: a more stabilized active preparation. *Biocatal Biotransform.*, 2007, 25, 453–458.

[15] Neta, JLV; DaSilva, LA; Lima, AA; Santana, JC; Leite, NS; Ruzene, DS; Silva, DP; DeSouza, RR. Bromelain enzyme from pineapple: *In*

vitro activity study under different micropropagation conditions. *Appl Biochem Biotechnol.*, 2012, 168, 234-246.

[16] Koh, J; Kang, SM; Kim, SJ; Cha, MK; Kwon, YJ. Effect of pineapple protease on the characteristics of protein fibers. *Fibers Polym.*, 2006, 7, 180–185.

[17] Orsini, RA. Bromelain. *Plast Reconstr Surg.*, 2006, 118, 1640-1644.

[18] Scopes, RK. *Protein Purification: Principles and Practice*, 2nd edition. New York: Springer-Verlag, 1982, p. 41-65.

[19] Rathnavelu, V; Alitheen, NB; Subramaniam, S; Samikannu, K; Ramesh, R. Potential role of bromelain in clinical and therapeutic applications. *Biomed Rep.*, 2016, 5, 283–288.

[20] Errasti, ME; Prospitti, A; Viana, CA; Gonzalez, MM; Ramos, MV; Rotelli, AE; Caffini, NO. Effects on fibrinogen, fibrin, and blood coagulation of proteolytic extracts from fruits of *Pseudananas macrodontes*, *Bromeliabalansae*, and *B. hieronymi* (Bromeliaceae) in comparison with bromelain. *Blood Coagul Fibrinolysis*. Epub, 2016, 27, 441-449.

[21] MacKay, ND; Miller, AL. Nutritional support for wound healing. *Altern Med Rev.*, 2003, 8, 359-377.

[22] Dhandayuthapani, S; Perez, HD; Paroulek, A; Chinnakkannu, P; Kandalam, U; Jaffe, M; Rathinavelu, A. Bromelain-induced apoptosis in GI-101A breast cancer cells. *J Med Food.*, 2012, 15, 344-349.

[23] Brien, S; Lewith, G; Walker, A; Hicks, SM; Middleton, D. Bromelain as a treatment for osteoarthritis: a review of clinical studies. *Evidence-based Complement Alternative Med.*, 2004, 1, 251-257.

[24] Maurer, HR. Bromelain: biochemistry, pharmacology and medical use. *Cell Mol Life Sci.*, 2001, 58, 1234-1245.

[25] Ota, S; Horie, K; Hagino, F; Hashimoto, C; Date, H. Fractionation and some properties of the proteolytically active components of bromelains in the stem and the fruit of the pineapple plant. *J Biochem.*, 1972, 71, 817-830.

[26] Miller, JM; Jr. Miller, JM; Opher, AW. (1970). The isoenzymes of bromelain. *Exp. Med. Surg.*, 28, 270-273.

[27] Ota, S; Muta, E; Katanita, Y; Okamoto, Y. Reinvestigation of fractionation and some properties of the proteolytically active components of stem and fruit bromelains. *J Biochem.*, 1985, 98, 219-228.

[28] Harrach, T; Eckert, K; Schulze-Forster, K; Nuck, R; Grunow, D; Maurer, HR. Isolation and partial characterisation of basic proteinases from stem bromelain. *J Prot Chem*, 1995, 14, 41-52.

[29] Lopez-Garcia, B; Hernandez, M; Segundo, BS. Bromelain, a cysteine protease from pineapple (*Ananas comosus*) stem, is an inhibitor of fungal plant pathogens. *Lett Appl Microbiol.*, 2012, 55, 62–67.

[30] Cohen, LW; Coghlan, VM; Dihel, LC. Cloning and sequencing of papain-encoding cDNA. *Gene.*, 1986, 48, 21–27.

[31] Chakrabarti, C; Biswas, S; Kundu, S; Sundd, M; Jagannadham, MV; Dattagupta, JK. Crystallization and preliminary X-ray analysis of ervatamin B and C, two thiol proteases from *Ervatamia coronaria. Acta Crystallogr D Biol Crystallogr.*, 1999, 55, 1074–1075.

[32] Sharma, K; Bhattacharyya, D. Application of transverse urea gradient zymography for structural and functional characterization of proteolytic enzymes. *Curr Sci.*, 2016, 111, 1340-1348.

[33] Carne, A; Moore, CH. The amino acid sequence of the tryptic peptides from actinidin, a proteolytic enzyme from the fruit of *Actinidia chinensis*. *Biochem J.*, 1978, 173, 73–83.

[34] Kamphuis, IG; Kalk, KH; Swarte, MBA; Drenth, J. Structure of papain refined at 1.65 Å resolution. *J Mol Biol.*, 1984, 179, 233–256.

[35] Ritonja, A; Rowan, AD; Buttle, DJ; Rawlings, ND; Turk, V; Barrett, AJ. Stem bromelain: amino acid sequence and implications for weak binding of cystatin. *FEBS Lett.*, 1989, 247, 419-424.

[36] Theodorou, LG; Bieth, JG; Papamichael, EM. The catalytic mode of cysteine proteinases of papain (C1) family. *Bioresour Technol.*, 2007, 98, 1931-1939.

[37] http://www.uniprot.org/uniprot/O23791.

[38] Ako, H; Cheung, AHS; Matsuura, PK. Isolation of a fibrinolysis enzyme activator from commercial bromelain. *Arch. Int. Pharmacodynamics*., 1981, 254, 157-167.

[39] Bhattacharyya, BK. Bromelain: An overview. *Nat. Prod. Radiance*., 2008, 7, 359-363.

[40] Hauck, JC; Chang, CM; Klein, G. Isolation of an effective debriding agent from the stems of pineapple plants. *Int J Tissue React*., 1983, 5, 125-134.

[41] Klaue, P; Dilbert, G; Hinke, G. *Tierexperimentelle Untersuchungen zur enzymatischen Lokalbenhandlung subderma ler Verbrennungen mit Bromelain Therapiewoche*., [*Animal studies on the enzymatic local treatment of subdermal burns with Bromelain Therapiewoche*.,] 1979, 29, 796-799

[42] Drenth, J; Jansonius, JN; Koekoek, R; Swen, HM; Wolters, BG. Structure of papain. *Nature*, 1968, 218, 929-932.

[43] PDB ID, 1IWD. Biswas, S; Chakrabarti, C; Kundu, S; Jagannadham, MV; Dattagupta, JK. Proposed amino acid sequence and the 1.63 A X-ray crystal structure of a plant cysteine protease, ervatamin B: some insights into the structural basis of its stability and substrate specificity. *Proteins*, 2003, 51, 489-497.

[44] PDB ID: 1W0Q. Banerjee, N; Das, S; Banerjee, A. A Three Dimensional Structural Model Of Stem Bromelain: A Critical Comparative Analysis Of Glycosylation Towards Functional And Structural Stability. (Unpublished data).

[45] Larocca, M; Rossano, R; Santamaria, M; Riccio, P. Analysis of pineapple [*Ananas comosus (L.) Merr.*] fruit proteinases by 2-D zymography and direct identification of the major zymographic spots by mass spectrometry. *Food Chem*., 2010, 123, 1334-1342.

[46] Corzo, CA; Krzysztof, Waliszewski KN; Welti-Chanes, J. Pineapple fruit bromelain affinity to different protein substrates. *Food Chem*., 2011, 133, 631-635.

[47] Illanes, A. Enzyme Production. In: *Enzyme Biocatalysis*. Brazil: Springer Science and Business Media V.B., 2008, p. 57-106.

[48] Hung, TH; Chang, YM; Sung, HY; Chang, CT. Purification and characterization of hydrolase with chitinase and chitosanase activity from commercial stem bromelain. *J Agric Food Chem.*, 2002, 50, 4666-4673.

[49] Bresolin, IRAP; Bresolin, ITL; Silveira, E; Tambourgi, EB; Mazzola, PG. Isolation and purification of bromelain from waste peel of pineapple for therapeutic application. *Braz Arch Biol Technol.*, 2013, 56, 971-979.

[50] Swaroop, G; Viswanathan, G. Isolation and characterization of bromelain (BML) proteases from *Ananas comosus:* an asset to cancer chemotherapy. *Int J Pharmacol Toxicol.*, 2013, 1, 82-90.

[51] Arumugam, A; Ponnusami, V. Pineapple fruit bromelain recovery using recyclable functionalized ordered mesoporous silica synthesized from sugarcane leaf ash. *Braz J Chem Eng.*, 2013, 30, 477-486.

[52] Yin, L; Sun, CK; Han, X; Xu, L; Xu, Y; Qi, Y; Peng, J. Preparative purification of bromelain (EC 3.4.22.33) from pineapple fruit by high-speed counter-current chromatography using a reverse-micelle solvent system. *Food Chem.*, 2011, 129, 925-932.

[53] Bhattacharya, R; Bhattacharyya, D. Resistance of Bromelain to SDS Binding. *Biochim Biophys Acta.*, 2009, 1794, 698-708.

[54] Chen, D; Huang, S. Fast separation of bromelain by polyacrylic acid-bound iron oxide magnetic nanoparticles. *Process Biochem.*, 2004, 39, 2207-2211.

[55] Babu, BR; Rastogi, NK; Raghavarao, KSMS. Liquid-liquid extraction of bromelain and polyphenol oxidase using aqueous two-phase system. *Chem Eng Process.*, 2008, 47, 83-89.

[56] Chen, T; Su, S; Nie, H; Zhu, L. Display: Design and selection of functional binding peptides ligands application for isolation of bromelain using affinity chromatography. *J Biotechnol.*, 2008, 136, 620-632.

[57] Devi, A. Sowmiya. Extraction and Characterization of Bromelain Enzyme from Pineapple (*Ananas comosus*) - Analysis of its Anti

Browning Activity. *J International Academic Research for Multidisciplinary*, 2014, 2, 284-295.

[58] Ng, PK; He, J; Synder, MK. Separation of proteins mixtures using pH-gradient cation exchange chromatography. *J Chromatogr A.*, 2009, 1216, 1372-1376.

[59] Nie, H; Li, S; Zhou, Y; Chen, T; He, Z; Su, S; Zhang, H; Xue, Y; Zhu, L. Purification of bromelain using immobilized metal affinity membranes. *J Biotechnol.*, 2008, 136, 620-632.

[60] Silveira, E. Expanded Bed Adsorption of Bromelain (E.C. 3.4.22.33) from *Ananas comosus* Crude Extract. *Brazil J Chem Eng.*, 2009, 26, 149-157.

[61] Cassano, A; Drioli, E; Galalaverna, G; Marchelli, R; Silvestro, R; Cagnasso, P. Clarification and concentration of citrus and carrot juices by integrated membrane processes. *J Food Eng.*, 2003,57, 153-163.

[62] Doko, B; Bassani, V; Casadebaig, J; Cavailles, L; Jacob, M. Preparation of proteolytic enzyme extracts from *Ananas comosus L. Merr.* fruit juice using semi permeable membrane, ammonium sulfate extraction, centrifugation and freeze-drying processes. *J Immunopharmaco.*, 2005, 4, 783-795.

[63] Lopes, FLG; Junior, JBS; Souza, RR; Ehrhardt, DD; Santana, JCC; Tambourgi, EB. Concentration by Membrane Separation Processes of a Medicinal Product Obtained from Pineapple Pulp. *Braz Arch Biol Technol.*, 2009, 52, 457-464.

[64] Saxena, L; Iyer, BK; Ananthanarayan, L. Three phase partitioning as a novel method for purification of ragi (*Eleusinecoracana*) bifunctional amylase/ protease inhibitor. *Process Biochem.*, 2007, 42,491-495.

[65] Silvestre, MPC; Carreira, RL; Silva, MR; Corgosinho, FC; Monteiro, MRP; Morais, HA. Effect of pH and temperature on the activity of enzymatic extracts from pineapple peel. *Food Bioprocess Technol.*, 2012, 5,1824-1831.

[66] Soares, PAG; Vaz, AFM; Correia, MTS; Pessoa, A; Maria, G; Cunha, C. Purification of bromelain from pineapple wastes by ethanol precipitation. *Sep Purif Technol.*, 2012, 98,389-395.

[67] Johansson, HO; Feitosa, E; Pessoa, A. Phase Diagrams of the Aqueous Two-Phase Systems of Poly (ethylene glycol)/Sodium Polyacrylate/Salts. *Polymers.*, 2011, 3,587-601.

[68] Rabelo, APB; Tambourgi, EB; Pessoa, A. Bromelain partioning in two phase aqueous systems containing PEO-PPO-PEO block copolymers. *J Chromatogr B.*, 2004, 807,61-68.

[69] Kaur, H; Corscadden, K; Lott, C; Elbatarny, HS; Othman, M. Bromelain has paradoxical effects on blood coagulability: a study using thromboelastography. *Blood Coagul Fibrinolysis.*, 2016, 27,745-752.

[70] Ketnawa, S; Chaiwut, P; Rawdkuen, S. Aqueous Two-phase Extraction of Bromelain from Pineapple Peels ('PhuLae' cultv.) and Its Biochemical Properties. *Food Sci Biotechnol.*, 2011, 20,1219-1226.

[71] Andre, A; Ninham, BW; Pileni, MP. *New J Chem.*, 2001, 25,563-571.

[72] Lee, KL; Chong, CC. *The use of reverse micelles in downstream processing of biotechnological products*. 2011, E-version.

[73] Chen, T; Su, S; Nie, H; Zhu, L. Display: Design and selection of functional binding peptides ligands application for isolation of bromelain using affinity chromatography. *J Biotechnol.*, 2008, 136,620-632.

[74] Bhattacharya, R. *Fruit and stem bromelain From pineapple (Ananas comosus): Stabilization and biochemical characterization of the enzymes*. Ph.D thesis, Jadavpur University, Kolkata, India., 2009.

[75] Bobb. Isolation of stem bromelain by affinity chromatography and its partial characterisation by gel electrophoresis. *Prep Biochem.*, 1972, 2,347-354.

[76] Murachi, T; Yamazaki, M. Changes in conformation and enzyme activity of stem bromelain in alkaline media. *Biochemistry.*, 1970, 9,1935-1938.

[77] Singh, LR; Devi, TP; Devi, SK. Purification and characterisation of a pineapple crown leaf thiol protease. *Prep Biochem Biotech.*, 2004, 34,25-43.

[78] Hale, LP; Greer, PK; Trinh, CT; James, CL. Proteinase activity and stability of natural bromelain preparations. *Int. Immunopharmacol.*, 2005, 5,783-793.

[79] Pardo, MF; Lopez, LM; Canals, F; Aviles, FX; Natalucci, CL; Caffini, NO. Purification of balansain I; an endopeptidase from unripe fruits of *Bromelia balansae* Mez (Bromeliaceae). *J Agr Food Chem.*, 2000, 48,3795-3800.

[80] Valles, D; Furtado, S; Cantera, AMB. Characterization of news proteolytic enzymes from ripe fruits of *Bromelia antiacantha* Bertol. (Bromeliaceae). *Enzyme Microb Tech.*, 2007, 40,409-416.

[81] Chaurasiya, RS; Umesh Hebbar, H. Extraction of bromelain from pineapple core and purification by RME and precipitation methods. *Sep Purif Technol.*, 2013, 111,90–97.

[82] Anwar, T; Ahmad, B; Younus, H. Cross-linked stem bromelain: a more stabilized active preparation. *Biocatal Biotransform.*, 2007, 25,453–458.

[83] Bhattacharya, R; Bhattacharyya, D. Preservation of natural stability of fruit "Bromelain" from *Ananas Comosus* (pineapple). *J Food Biochem.*, 2009, 33,1-19.

[84] Frazier, WC. Preservation of fruits and fruit products. In: *Food Microbiology*. New York: Mcgraw-Hill Book Company, Inc., 1958, 177-181.

[85] Mynott, TL; Luke, RK; Chandler, DS. Oral administration of protease inhibits enterotoxigenic *Escherichia coli* receptor activity in piglet small intestine. *Gut.*, 1996, 38,28-32.

[86] Mynott, TL; Guandalini, S; Raimondi, F; Fasano, A. Bromelain prevents secretion caused by Vibrio cholerae and Escherichia coli enterotoxins in rabbit ileum *in vitro*. *Gastroenterology*, 1997, 113,175-181.

[87] Falanga, V. Wound bed preparation and the role of enzymes: A case for multiple actions of therapeutic agents. *Wounds*, 2002, 14,47-57.

[88] Gupta, P; Saleemuddin, M. Bioaffinity Based Oriented immobilization of stem bromelain. *Biotechnol Lett.*, 2006, 28,917–922.

[89] Cunningham, EL; Jaswal, SS; Sohl, JL; Agard, DA. Kinetic stability as a mechanism for protease longevity. *Proc Natl Acad Sci.*, 1996, 96,11008-11014.

[90] Jaswal, SS; Truhlar, SM; Dill, KA; Agard, DA. Comprehensive analysis of protein folding activation thermodynamics reveals a universal behavior violated by kinetically stable proteases. *J Mol Biol.*, 2005, 347,355-366.

[91] Manning, M; Colon, W. Structural basis of protein kinetic stability: resistance to sodium dodecyl sulfate suggests a central role for rigidity and a bias toward beta-sheet structure. *Biochemistry*, 2004, 43,11248-11254.

[92] Nishikori, S; Shiraki, K; Okanojo, M; Imanaka, T; Takagi, M. Equilibrium and kinetic stability of a hyperthermophilic protein, O^6-methylguanine-DNA methyltransferase under various extreme conditions.*J Biochem.*, 2004, 136,503-508.

[93] DeLong, EF. A phylogenetic perspective on hyperthermophilic microorgansims. *Meth Enzymol.*, 2001, 330,3-11.

[94] Huber, R; Stetter, KO. Discovery of hyperthermophilic microorganisms. *Meth. Enzymol.*, 2001, 330,11-24.

[95] Szilagyi, A; Zavodszky, P. Structural differences between mesophilic: moderately thermophilic and extremely thermophilic protein subunits: results of a comprehensive survey. *Structure.*, 2000, 8,493-504.

[96] Fan, KK; Ouyang, P; Wu, X; Lu, Z. A model of interfacial inactivation for papain in aqueous organic biphasic systems. *Enzyme Microb Technol.*, 2001,8,3-7.

[97] Lou, WY; Zong, MH; Smith, TJ; Wu, H; Wang, JF. Impact of ionic liquids on papin, an investigation of structure-function relationships. *Green Chem.*, 2006, 8,509-512.

[98] Ness, JE; Welch, M; Giver, L; Bueno, M; Cherry, JR; Borchert, TV; Stemmer, WP; Minshull, J. DNA shuffling of genomic sequences of subtilisin. *Nat Biotechnol.*, 1999, 17,893-896.

[99] Szabo, A; Kotorman, M; Laczko, I; Simona, LM. Spectroscopic studies of stability of papain in aqueous organic solvents. *J Mol Catal B: Enzyme.*, 2006, 41,43-48.

[100] Zhao, H; Arnold, FH. Directed evolution subtilisin E into a functional equivalent of thermitase. *Protein Eng.*, 1999, 12,47-53.

[101] Matthews, BW. Studies on protein stability with T4 lysozyme. *Adv. Protein Chem.*, 1995, 46,249-278.

[102] Ahmad, B; Shamim, TA; Haq, SK; Khan, RH. Identification and characterization of functional intermediates of stem bromelain during urea and guanidine hydrochloride unfolding. *J Biochem.*, 2007, 141,251-259.

[103] Arroyo-Reyna A; Hernández-Arana A. The thermal denaturation of stem bromelain is consistent with an irreversible two-state model. *Biochim Biophys Acta.* 1995, 1248, 123-128.

[104] Gupta P; Khan RH; Saleemuddin M. Trifluoroethanol-induced "molten globule" state in stem bromelain. *Arch Biochem Biophys.*, 2003, 413,199-206.

[105] Haq, SK; Rasheedi, S; Khan, RH. Characterization of a partially folded intermediate of stem bromelain at low pH. *Eur J Biochem.*, 2002, 269, 47-52.

[106] Haq, SK; Rasheedi, S; Sharma, P; Ahmad, B; Khan, RH. Influence of salts and alcohols on the conformation of partially folded intermediate of stem bromelain at low pH. *Int J Biochem Cell Biol.*, 2005, 37,361-374.

[107] Khan, RH; Rasheedi, S; Haq, SK. Effect of pH, temperature and alcohols on the stability of glycosylated and deglycosylated stem bromelain. *J Biosci.*, 2003, 6, 709-714.

[108] Rasheedi, S; Haq, SK; Khan, RH. Guanidine hydrochloride denaturation of glycosylated and deglycosylated stem bromelain. *Biochemistry (Moscow)*, 2003, 68, 1097-1100.

[109] Creighton, TE. Electrophoretic analysis of the unfolding of proteins by urea. *J Mol Biol.*, 1979, 129,235–264.
[110] Creighton, TE. Kinetic study of protein unfolding and refolding using urea gradient electrophoresis. *J Mol Biol.*, 1980, 137,61–80.
[111] Gentile, F; Veneziani, BM; Sellitto, C. Polyacrylamide gel electrophoresis in discontinuous transverse urea-gradient gels. *Anal Biochem.*, 1997, 244,228–232.
[112] Goldenberg, DP; Creighton, TE. Gel electrophoresis in studies of protein conformation and folding. *Anal Biochem.*, 1984, 138,1–18.
[113] Goldenberg, DP. Analysis of protein conformation by gel electrophoresis. In *Protein Structure: A Practical Approach* (ed. Creighton TE.), IRL Press, UK: Oxford., 1989, 225–250.
[114] Creighton, TE. Protein folding. *Biochem J.*, 1990, 270,1–16.
[115] Waner, MJ; Navrotskaya, I; Bain, A; Oldham, ED; Mascotti, DP. Thermal and sodium dodecylsulfate induced transitions of streptavidin. *Biophys J.*, 2004, 87,2701-2713.
[116] Tsou, CL. Inactivation precedes overall molecular conformation changes during enzyme denaturation. *Biochim Biophys Acta.*, 1995, 1253,151 – 162.
[117] Jaenicke, R; Böhm, G. The stability of proteins in extreme environments. *Curr Opin Struct Biol.*, 1998, 8,738-748.
[118] Jaenicke, R. Protein stability and molecular adaptation to extreme conditions. *Eur J Biochem.*, 1991, 202,715-728.
[119] Rees, DC; Adams, MWW. Hyperthermophiles: taking the heat and loving it. *Structure*, 1995, 3,251-254.
[120] Jaenicke, R. Do ultrastable proteins from hyperthermophiles have high or low conformational rigidity? *Proc Natl Acad Sci.*, 2000, 97,2962-2964.
[121] Razvi, A; Scholtz, JM. Lessons in stability from thermophilic proteins. *Protein Sci.*, 2006, 15,1569-78.
[122] Rees, DC; Robertson, AD. Some thermodynamic implications for the thermostability of proteins. *Protein Sci.*, 2001, 10,1187-1194.

[123] Brito, RM; Vaz, WL. Determination of the critical micelle concentration of surfactants using the fluorescent probe N-phenyl-1-naphthylamine. *Anal Biochem.*, 1986, 152,250-255.

[124] Bhattacharya, R; Mukherjee, D; Bhattacharyya, D. Quantification of protein-bound sodium dodecyl sulfate by Rhodamine B: A method for identification of kinetically stable proteins. *Anal Biochem.*, 2011, 417,17-24.

[125] La Mesa, C. Polymer-surfactant and protein-surfactant interactions. *J Colloid Interface Sci.*, 2005, 286,148-157.

[126] Jones, MN. Surfactant interactions with biomembranes and proteins. *Chem Soc Rev.*, 1992, 21,127-136.

[127] Jones, MN; Manley, P. Binding of n-alkyl sulphates to lysozyme in aqueous solution. *J. Chem Soc Faraday Trans.*, 1979, 75, 1736-1744.

[128] Reynolds, JA; Herbert, S; Polet, H; Steinhardt, J. The binding of diverse detergent anions to bovine serum albumin. *Biochemistry*, 1967, 6,937-943.

[129] Helenius, A; Simons, K. Solubilization of membranes by detergents. *Biochim Biophys Acta.*, 1975, 415,29-79.

[130] Durchschlag, H; Weber, R; Jaenicke, R. In: *Proceedings of the 4th World Surfactants Congress*. *CESIO*,Barcelona, 1996, 1,519-534.

[131] Jirgensons, B. Conformational transitions of non-helical proteins effected by dodecyl sulfate. Circular dichorism of alpha1-acid glycoprotein, Bence Jones protein, carbonic anhydrase B, deoxyribonuclease, pepsinogen and plasminogen. *Biochim Biophys Acta.*, 1976, 434,58-68.

[132] Hale, LP. Proteolytic activity and immunogenicity of oral bromelain within the gastrointestinal tract of mice. *Int Immunopharmacol.*, 2004, 4,255-264.

[133] Hale, LP; Greer, PK; Trinh, CT; James, CL. Proteinase activity and stability of natural bromelain preparations. *Int Immunopharmacol.*, 2005, 5,783-793.

[134] Hale, LP; Greer, PK; Sempowski, GD. Bromelain treatment alters leukocyte expression of cell surface molecules involved in cellular adhesion and activation. *Clin Immunol.*, 2002, 104,183 – 190.

[135] Castell, JV; Friedrich, G; Kuhn, CS; Poppe, GE. Intestinal absorption of undegraded proteins in men: presence of bromelain in plasma after oral intake. *Am J Physiol.*, 1997, 273,139 – 146.

[136] Dutta, S. *Evaluation and biosynthesis of enzymes from tropical medicinal plants.* PhD thesis, Jadavpur University, Kolkata, India. 2013.

[137] Pirotta, F; de Giuli-Morghen, C. Bromelain: anti-inflammatory and serum fibrinolytic activity after oral administration in the rat. *Drugs Exp Clin Res.*, 1978, 4, 1–20.

[138] Manzoor, Z; Nawaz, A; Mukhtar, H; Haq, I. Bromelain: Methods of Extraction, Purification and Therapeutic Applications. *Brazilian Arch Biol Technol.*, 2016, 59, e16150010.

[139] Livio, M; Gaetano, GD; Donati, MB. Effect of bromelain of fibrinogen level, protrombin complex and platelet aggregation in the rat - a preliminary report. *Drugs under Exp. Clin Res.*, 1978, 1, 49–53.

[140] Morita, AH; Uchida, DA; Taussig, SJ. Chromatographic fractionation and characterization of the active platelet aggregation inhibitory factor from bromelain. *Arch Int Pharmacodyn Ther.*, 1979, 239, 340-350.

[141] Barth, H; Guseo, A; Klein, R. *In vitro* study on the immunological effect of bromelain and trypsin on mononuclear cells from humans. *Eur J Med Res.*, 2005, 10, 325–331.

[142] Hale, LP; Haynes, BF. Bromelain treatment of human T cells removes CD44, CD45RA, E2/MIC2, CD6, CD7, CD8, and Leu 8/LAM1 surface molecules and markedly enhances CD2-mediated T cell activation. *J Immunol.*, 1992, 149, 3809–3816.

[143] Lehmann, PV. Immunomodulation by proteolytic enzymes. *Nephrol Dial Transplantation.*, 1996, 11, 953–955.

[144] Desser, L; Rehberger, A; Kokron, E; Paukovits, W. Cytokine synthesis in human peripheral blood mononuclear cells after oral administration of polyenzyme preparations. *Oncology.*, 1993, 50, 403–407.

[145] Desser, L; Rehberger, A; Paukovits, W. Proteolytic enzymes and amylase induce cytokine production in human peripheral blood mononuclear cells *in vitro. Cancer Biotherapy.*, 1994, 9, 253–263.

[146] Eckert, K; Grabowska, E; Stange, R; Schneider, U; Eschmann, K; Maurer, HR. Effects of oral bromelain administration on the impaired immunocytotoxicity of mononuclear cells from mammary tumor patients. *Oncology Reports.*, 1999, 6, 1191–1199.

[147] Engwerda, CR; Andrew, D; Murphy, M; Mynott, TL. Bromelain activates murine macrophages and natural killer cells *in vitro.Cell Immunol.*, 2001, 210, 5–10.

[148] Engwerda, CR; Andrew, D; Ladhams, A; Mynott, TL. Bromelain modulates T cell and B cell immune responses *in vitro* and *in vivo. Cell Immunol.*, 2001, 210, 66–75.

[149] Mynott, TL; Ladhams, A; Scarmato, P; Engwerda, CR. Bromelain, from pineapple stems, proteolytically blocks activation of extracellular regulated kinase-2 in T cells. *J Immunol.*, 1999, 163, 2568–2575.

[150] Secor, Jr. ER; Singh, A; Guernsey, LA; McNamara, JT; Zhan, L; Maulik, N; Thrall, RS. Bromelain treatment reduces CD25 expression on activated CD4+ T cells *in vitro. Int Immunopharmacol.*, 2009, 9, 340–346.

[151] Leipner, J; Iten, F; Saller, R. Therapy with proteolytic enzymes in rheumatic disorders. *BioDrugs.*, 2002, 15, 779–789.

[152] Chobotova, K; Vernallis, AB; Majid, FAA. Bromelain's activity and potential as an anti-cancer agent: Current evidence and perspectives. *Cancer Lett.*, 2010, 290, 148-156.

[153] Beez, R; Lopes, MTP; Salas, CE; Hern´andez, M. *In vivo* antitumoral activity of stem pineapple (Ananas comosus) bromelain. *Planta Medica.*, 2007, 73, 1377–1383.

[154] Taussig, SJ; Szekerczes, J; Batkin, S. Inhibition of tumour growth *in vitro* by bromelain, an extract of the pineapple plant (Ananas comosus). *Planta Medica.*, 1985, 6, 538–539.

[155] Tysnes, BB; Maurer, HR; Porwol, T; Probst, Bjerkvig R; Hoover, F. Bromelain reversibly inhibits invasive properties of glioma cells. *Neoplasia.*, 2001, 3, 469–479.

[156] Mantovani, A; Allavena, P; Sica, A; Balkwill, F. Cancer related inflammation. *Nature*, 2008, 454, 436–444.

[157] Ferris, RL; Grandis, JR. NF-κB gene signatures and p53 mutations in head and neck squamous cell carcinoma. *Clin Cancer Res.*, 2007, 13, 5663–5664.

[158] Hussain, SP; Harris, CC. Inflammation and cancer: an ancient link with novel potentials. *Int J Cancer.*, 2007, 121, 2373–2380.

[159] Wang, MT; Honn, KV; Nie, D. Cyclooxygenases, prostanoids, and tumor progression. *Cancer Metastasis Rev.*, 2007, 26, 525–534.

[160] Bhui, K; Prasad, S; George, J; Shukla, Y. Bromelain inhibits COX-2 expression by blocking the activation of MAPK regulated NF-kappa B against skin tumor-initiation 6 Biotechnology Research International triggering mitochondrial death pathway. *Cancer Lett.*, 2009, 282, 167–176.

[161] Huang, JR; Wu, CC; Hou, RCW; Jeng, KC. Bromelain inhibits lipopolysaccharide-induced cytokine production in human THP-1 monocytes via the removal of CD14. *Immunol Invest.*, 2008, 37, 263–277.

[162] Hou, RCW; Chen, YS; Huang, JR; Jeng, KCG. Cross-linked bromelain inhibits lipopolysaccharide-induced cytokine production involving cellular signaling suppression in rats. *J Agri. Food Chem.*, 2006, 54, 2193–2198.

[163] Lawrence, RC; Helmick, CG; Arnett, FC; Deyo, RA; Felson, DT; Giannini, EH; Heyse, SP; Hirsch, R; Hochberg, MC; Hunder, GG; Liang, MH; Pillemer, SR; Steen, VD; Wolfe, F. Estimates of prevalence of arthritis and selected musculoskeletal disorders in the United States. *Arthritis Rheum.*, 1998, 41, 778–799.

[164] Akhtar, NM; Naseer, R; Farooqi, AZ; Aziz, W; Nazir, M. Oral enzyme combination versus diclofenac in the treatment of osteoarthritis of the knee—a double-blind prospective randomized study. *Clin Rheumatology*, 2004, 23, 410–415.

[165] Brien, S; Lewith, G; Walker, A; Hicks, SM; Middleton, D. Bromelain as a treatment for osteoarthritis: a review of clinical studies. *Evidence-Based Complement Alt Med.*, 2004, 1, 251–257.

[166] Mojcik, CF; Shevach, EM. Adhesion molecules: a rheumatologic perspective. *Arthritis Rheum.*, 1997, 40, 991–1004.

[167] Bodi, T. The effects of oral bromelains on tissue permeability to antibiotics and pain response to bradykinin: double blind studies on human subjects. *Clin Med.*, 1966, 73, 61–65.

[168] Kumakura, S; Yamashita, M; Tsurufuji, S. Effect of bromelain on kaolin-induced inflammation in rats. *Eur J Pharmacol.*, 1988, 150, 295–301.

[169] Conrozier, T; Mathieu, P; Bonjean, M; Marc, JF; Renevier, JL; Balblanc, JC. A complex of three natural anti-inflammatory agents provides relief of osteoarthritis pain. *Altern Ther Health Med.*, 2014, 20, 32-37.

[170] Moss, JN; Frazier, CV; Martin, GJ. Bromelains: The pharmacology of the enzymes. *Arch Int Pharmacodyn.*, 1963, 145, 166-189.

[171] Friesen, A; Schilling, A; Hofstetter, A; Adam, D. Tetracyclin-Konzentration im Prostata-Sekret. [Tetracycline concentration in the prostatic secretion] *Z antimikrob antineoplast Chirurgie.*, 1987, 2, 61–65.

[172] Renzini, G; Varengo, M. Die Resorption von Tetrazyklinin Gegenwart von Bromelain bei oraler Applikation. [The absorption of tetracycline in the presence of bromelain in oral administration.] Arzneimittel-Forsch. *Drug Res.*, 1972, 2, 410–412.

[173] Tinozzi, S; Venegoni, A. Effect of bromelain on serum and tissue levels of amoxycillin. *Drug Exp Clin Res.*, 1978, 1, 39–44.

[174] Neubauer, RA. A plant protease for potentiation of and possible replacement of antibiotics. *Exp Med Surg.*, 1961, 19, 143–160.

[175] Guo, R; Canter, PH; Ernst, E. Herbal medicines for the treatment of rhinosinusitis: a systematic review, Otolaryn. *Head Neck Surg.*, 2006, 135, 496-506.

[176] Ryan, RE. A double- blind clinical evaluation of bromelain in the treatment of acute sinusitis. *Headache*, 1967, 4, 13–17.

[177] Helms, S; Miller, A. Natural treatment of chronic rhinosinusitis. *Altern Med Rev.*, 2006, 11, 196-207.

[178] Tochi, BN; Wang, Z; Xu, SY; Zhang, W. Therapeutic application of pineapple protease (Bromelain): A review. *Pakistan J Nutrition*, 2008, 7, 513-520.

[179] Singer, AJ; McClain, SA; Taira, BR; Rooney, J; Steinhauff, N; Rosenberg, L. Rapid and selective enzymatic debridement of porcine comb burns with bromelain-derived Debrase: acute-phase preservation of noninjured tissue and zone of stasis. *J Burn Care Res.*, 2010, 31, 304-309.

[180] Ahle, NW; Hamlet, MP. Enzymatic frostbite eschar debridement by bromelain. *Ann Emerg Med.*, 1987, 16, 1063-1065.

[181] Singer, AJ; Taira, BR; Anderson, R; McClain, SA; Rosenberg, L. The effects of rapid enzymatic debridement of deep partial-thickness burns with Debrase on wound reepithelialization in swine. *J Burn Care Res.*, 2010, 31, 795-802.

[182] Hu, W; Wang, AM; Wu, SY; Zhang, B; Liu, S; Gou, YB; Wang, JM. Debriding effect of bromelain on firearm wounds in pigs. *J Trauma.*, 2011, 71, 966-972.

[183] Rosenberg, L; Krieger, Y; Silberstein, E; Arnon, O; Sinelnikov, IA; Bogdanov-Berezovsky, A; Singer, AJ. Selectivity of a bromelain based enzymatic debridement agent: a porcine study. *Burns*, 2012, 38, 1035-1040.

[184] Graf, J. Herbal anti-inflammatory agents for skin disease. *Skin Therapy Lett.*, 2000, 5, 3-5.

[185] Wu, SY; Hu, W; Zhang, B; Liu, S; Wang, JM; Wang, AM. Bromelain ameliorates the wound microenvironment and improves the healing of firearm wounds. *J Surgical Res.*, 2012, 176, 503–509.

[186] Mukherjee, D. *Multimeric proteins: Its adaptation and regulation of biological activities.* Ph.D thesis, Jadavpur University, Kolkata, India., 2013.

[187] Das, S; Bhattacharyya, D. Destabilization of human insulin fibrils by peptides of fruit bromelain derived from *Ananas comosus*

(pineapple). *J Cell Biochem.*, 2017. DOI: 10.1002/jcb.26173 [In Press].

[188] Dutta, S; Bhattacharyya, D. Enzymatic, antimicrobial and toxicity studies of the aqueous extract of *Ananas comosus* (pineapple) crown leaf. *J Ethnopharmacol.*, 2013, 150, 451-457.

[189] Taussig, SJ; Yokoyama, MM; Chinen, A. Bromelain: a proteolytic enzyme and its clinical application: a review. *Hiroshima J Med Sci.*, 1975, 24, 185–193.

In: Tropical Fruits
Editors: C. Stewart Bogsan et al.
ISBN: 978-1-53612-885-7

Chapter 4

FRESH-CUT PINEAPPLE: UV-C LIGHT POTENTIALITIES IN SHELF-LIFE EXTENSION

Stella Plazzotta [*] ***and Lara Manzocco***
Department of Agricultural, Food, Environmental and Animal Sciences, University of Udine, Udine, Italy

ABSTRACT

Consumer demand for fresh-cut fruit has substantially risen in the last years, due to high nutritional and sensory value of these products and to their high convenience. Pineapple is one of the most important fresh-cut fruit, being sold alone or mixed in fruit salads. However, it presents some major issues. In fact, it has a short shelf-life, being limited by microbial and enzymatic activity. Shelf-life extension of fresh-cut pineapple can be achieved by means of novel non-thermal technologies. Among them, UV-C light treatment seems to be very promising. It has been extensively used for surface decontamination of different food products including raw meat and fish, shell eggs, bakery products and fresh-cut fruits. In addition, different studies have assessed UV-C light

[*] Corresponding Author Email: stella.plazzotta@uniud.it.

treatment efficacy in the inactivation of fruit endogenous enzymes. Moreover, the technology is safe to apply when some simple precautions are adopted and it has not been reported to leave residues on treated food.

Based on these considerations, the aim of the present chapter is to present recent research outcomes in the field of UV-C light stabilization of fresh-cut pineapple. UV-C light treatment of fresh-cut pineapple is expected to be particularly challenging. This is due to the irregular surface of this fruit, which is responsible for the formation of shadowed zones, not reached by the UV-C radiation. The effect of packaging on UV-C light treatment efficacy is also discussed.

Keywords: fresh-cut, shelf-life, ultraviolet light, quality, packaging

INTRODUCTION

Fresh-cut products are fruits or vegetables that have been trimmed, peeled and/or cut into a fully usable product, which is subsequently packaged (IFPA, 2000). These products are fresh-like, healthy, convenient and additive-free foods which are increasingly demanded by consumers. Fresh-cut pineapple is one of the most popular fresh-cut products, being one of the most consumed fresh-cut fruit in US with apple and watermelon (Cook, 2015; Jennylynd & Ngarmsak, 2010). It is served in rings, sticks or cubes both alone and in mixed fruit formulations, giving consumers a broad choice. Typical fresh-cut pineapple processing is represented in Figure 1.

The careful selection of pineapple fruit, the disinfection in chlorinated water, often using antibrowning agents, and the maintenance of cold chain aim to preserve as long as possible safety and quality attributes of fresh-cut pineapple. Nevertheless, the shelf-life of this product is still limited to few days, due to enzymatic activity and microbial growth. The latter are associated with colour changes, tissue softening, juice leakage and off-odours development.

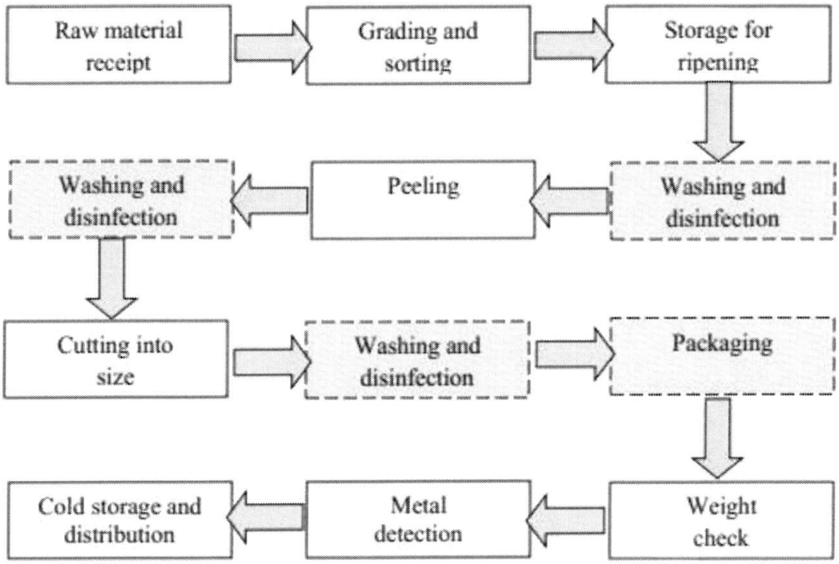

Figure 1. Typical fresh-cut pineapple processing. Unit operations in which UV-C light treatment could be applied are within dotted grey squares.

Fresh-cut processing is well known to promote a fast deterioration of fruit due to both environmental contamination and vegetable tissue damage (González-Aguilar, Ruiz-Cruz, Cruz-Valenzuela, Rodríguez-Félix, & Wang, 2004). For this reason, increasing attention has been payed to the development of additional strategies to prolong fresh-cut pineapple shelf-life. Among them, modified atmosphere packaging (Marrero & Kader, 2006), edible coatings (Azarakhsh, Osman, Ghazali, Tan, & Mohd Adzahan, 2012) and UV-C irradiation (Alothman, Bhat, & Karim, 2009) are some of the most promising.

UV-C Light Processing

Ultraviolet processing of solid food is U. S. FDA-approved (U. S. FDA, 2000) and does not require authorization according to EC novel food regulation (EC/258/97). UV-C light treatment exploits the radiation from the electromagnetic spectrum from 200 to 280 nm. The UV-C light technology is fast and economic, requiring a quite simple and cheap equipment, based on mercury lamps designed to produce energy in the germicidal region (254 nm). These lamps, emitting at a wavelength below 260 nm, produce ozone, which must be promptly removed to avoid operator health hazards (Bintsis, Litopoulou-Tzanetaki, & Robinson, 2000). The intensity of UV radiation is expressed as irradiance (I, W/m^2), which expresses the radiation intensity (P, W) on a given surface (S, m^2) (Eq. 1).

$$I = P / S \qquad \text{(Eq. 1)}$$

Irradiance is generally modified changing the distance between lamp and product surface and easily measured using a UV radiometer. Treatments are generally compared considering their light dose or fluence (F, J/m^2) that accounts for both radiation intensity (I) and exposure time (t, s) (Eq. 2) (Bintsis et al., 2000).

$$F = I \times t \qquad \text{(Eq. 2)}$$

UV-C treatment has been largely used in the food sector to decontaminate food surface, including raw meat, egg shells, bakery products and fresh-cut fruits. Regarding application to fresh-cut fruit, UV-C light has been mostly used as decontamination method, due to its ability to damage microbial DNA thorough dimer formation, leading to cell death (Sastry, Datta, & Worobo, 2000; Pan, Vicente, Martinez, Chaves, & Civello, 2004; Yaun, Sumner, Eifert, & Marcy, 2004). The application of UV-C treatments has not been reported to form known toxic byproducts (Keyser, Müller, Cilliers, Nell, & Gouws, 2008). Nevertheless, UV-C light was reported to modify fruit metabolism, antioxidant capacity and concentration of some bioactive (Freitas et al., 2015).

The efficacy of food treatment by UV-C light is strongly related to surface topography as well as to treatment conditions. In particular, an accurate 100% enlightening of the surface must be guaranteed since the germicidal effect of UV-C light is significantly impaired in shadowed zones. Similarly, if the food surface is non-homogenous and uneven, the UV-C effect will be impaired (Gardner & Shama, 2000). Products should be individually treated to prevent shadowing effects among different food items. The treatment is generally applied on unpackaged food to avoid the screening effect of packaging material and ensure maximum decontamination effectiveness (Fonseca & Rushing, 2006). However, the eventual occurrence of a post-treatment contamination is likely to nullify the UV-C decontamination efficacy. The use of a proper plastic material, transparent to UV-C light, could avoid this issue.

Based on these considerations, the application of UV-C light in a standard fresh-cut pineapple process could be applied both in disinfection unit operations and after packaging (Figure 1).

The aim of the present chapter was to evaluate the effectiveness of UV-C light in extending the shelf-life of fresh-cut pineapple sticks packaged in different conditions.

SURFACE DECONTAMINATION OF FRESH-CUT PINEAPPLE USING UV-C LIGHT

Conflicting results can be found in the literature about the effect of UV-C light treatment for shelf-life extension of fresh-cut pineapple, mostly due to different processing conditions. To this regard, positive and side effects of UV-C light treatments on fresh-cut pineapple described in some recent studies are reported in Table 1.

It must be observed that really different UV-C fluences have been applied in different studies (from 2 to 12000 J/m^2). Moreover, the same UV-C fluence can be obtained under completely different processing conditions. For instance, a dose of 1000 J/m^2 can be equally obtained by 2 min exposure to 8.5 W/m^2 UV light or by 1 min exposure to 17 W/m^2 UV light. Even though the UV dose is the same, the time scale of the phenomena affected by irradiation could significantly modify the overall efficiency of the process (Manzocco & Nicoli, 2015).

Table 1. Positive and side effects of UV-C light treatments on fresh-cut pineapple reported in some recent studies

UV-C light fluence (J/m^2)	Positive effects	Side effects	Reference
2	Increase of flavonoid content	Reduction of vitamin C content	Alothman, et al. (2009)
4500	Reduction of firmness loss Reduction of total solid content loss	Reduction of vitamin C content Increase of browning extent and rate	Pan & Zu (2012)
12000	Reduction of microbial surface contamination Reduction of juice leakage	Not reported	Manzocco, Da Pieve, Bartolomeoli, & Maifreni (2011)

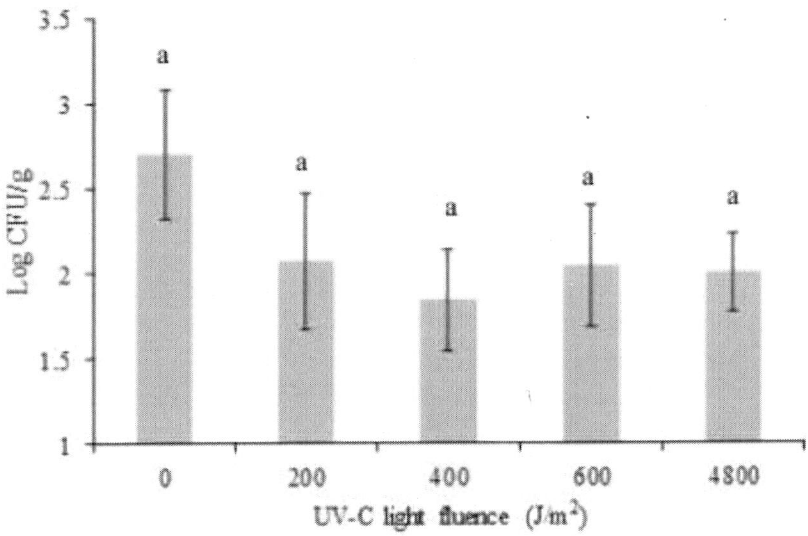

Figure 2. Yeast load of fresh-cut pineapple sticks submitted to increasing UV-C light fluences (0, 200, 400, 600, 4800 J/m^2). Modified from Manzocco et al. (2016).

In the case of pineapple, yeast cells are the decontamination target, as their growth is well known to be one of the most important factors limiting fresh-cut pineapple shelf-life (Chonhenchob, Chantarasomboon, & Singh, 2007; Raybaudi-Massilia, Mosqueda-Melgar, Soliva-Fortuny, & Martín-

Belloso, 2009). Recent data relevant to the efficacy of UV-C light in decontaminating fresh-cut pineapple sticks are reported in Figure 2. Pineapple sticks were treated at increasing UV-C light fluences (0, 200, 400, 600, 4800 J/m^2) and analysed for yeast load.

Collected data showed no significant yeast decontamination effect upon UV-C light treatment, even at the highest applied fluence. This result could be due to the critical pineapple fruit surface. The latter is in fact rough and uneven, leading to shadowed zones. Moreover, the peculiar fruit development as association of multiple fruitlets is well known to promote microbial internalization in hidden nests, not reached by the UV-C light (Rohrbach & Johnson, 2003). In the literature, higher fluences (12000 J/m^2) have been found to be effective in reducing fresh-cut pineapple surface contamination (Manzocco et al., 2011). However, the latter have also been associated to quality loss (Pan & Zu, 2012).

Effect of UV-C Light Treatment on Quality of Fresh-Cut Pineapple during Storage

Even if UV-C light at low fluences is not effective in decontaminating fresh-cut pineapple immediately after treatment, it can be able to stress yeast cells, leading to a reduced microbial growth during storage. A storage test can thus be helpful to better understand UV-C light treatment potentialities in extending the shelf-life of fresh-cut pineapple. In doing this, packaging conditions should be carefully selected.

The application of UV-C light treatment can be performed either before or after the packaging step (Figure 1). In the first case, any screening effect exerted by packaging material should be expected to hinder treatment efficacy. Nevertheless, treatment efficacy could be reduced by the occurrence of post-treatment contaminations. The latter are completely avoided if UV-C light treatment is applied on already packaged items. However, in such a case, the eventual reduction in UV-C light decontamination efficacy due to packaging screening should be evaluated.

For this reason, a highly UV-C light transparent packaging material should be used.

Following, recent data relevant to a storage test performed to assess UV-C light treatment efficacy in relation to packaging conditions are reported. In particular, fresh-cut pineapple sticks were submitted to 200 J/m^2 UV-C light irradiation, the treatment being applied either before and after packaging. In the former case a commonly used plastic PET tray was used; in the latter, a plastic PA/PE pouch transparent to UV-C light was used. Preliminarily, the UV-C light transmittance of the two packaging materials was assessed using a luminometer. Data confirmed that the PET tray completely screened the UV-C light, while the UV-C light transmittance of the PA/PE pouch resulted of 80%. Packaged pineapple sticks were stored in dark and refrigerated conditions for 15 days. During this time, different quality evaluations were performed, including determination of yeast growth, colour, head-space oxygen concentration and consumer preference. Results are reported in Table 2.

Data indicate that UV-C light treatment exerted a significant inhibitory effect on yeast growth. In fact, while in the control sample the microbial load reached 5 Logs after 7 days of refrigerated storage, UV-C light treatment, being applied either before and after packaging, was associated with a slower yeast growth, which did not exceed 4 Logs after 15 days.

This result confirms that even if UV-C light does not decrease microbial counts just after the treatment, it can exert inhibitory effects during the following product storage.

UV-C light treatment has been reported to possibly modify fruit respiration metabolism. To this regard Cai, Shi, Chen, & Yang (2015) reported that strawberry respiration rate significantly decreased upon a 3000 J/m^2 UV-C light treatment. Table 2 shows that respiration kinetics of pineapple tissue was not modified. Oxygen depletion in the packages resulted in fact similar in control and in UV-C light treated samples. In all samples, oxygen was completely depleted within 7 days, in agreement with results obtained by Chonhenchob et al. (2007). This is probably due to the lower fluence (200 J/m^2) as compared to that applied to strawberry (Cai et al., 2015).

Table 2. Yeast load, oxygen head-space concentration, colour and consumer preference evaluated during a 15-day refrigerated storage on untreated fresh-cut pineapple sticks (control) and on samples submitted to 200 J/m² UV-C light fluence before (UV-C pre-packaging) and after (UV-C post-packaging) packaging

Parameter		Sample	Time (days)				
			0	3	7	10	15
Yeasts (Log10 CFU/g)		Control	2.8±0.2 [b]	2.9±0.2 [b]	4.9±0.8 [a]	4.5±0.5 [a]	4.4±0.4 [a]
		UV-C pre-packaging	2.2±0.2 [b]	2.0±0.3 [b]	2.4±0.7 [b]	2.5±0.6 [b]	3.0±0.6 [ab]
		UV-C post-packaging	2.4±0.1 [b]	2.6±0.2 [b]	2.3±0.6 [b]	2.8±0.4 [ab]	3.4±0.6 [ab]
O_2 concentration (%)		Control	21.1±1.2	7.5±1.3	0.3±0.2	0.3±0.1	0.3±0.2
		UV-C pre-packaging	21.0±0.9	5.2±0.7	0.3±0.2	0.3±0.0	0.3±0.0
		UV-C post-packaging	21.0±0.9	5.6±0.8	0.3±0.1	0.3±0.2	0.3±0.2
Colour	L*	Control	73±2 [ab]	73±1[a]	74±1[a]	71±2 [b]	71±2 [b]
	a*		-3±1 [a]	-4±1 [a]	-4±1 [a]	-4±1 [a]	-4±1 [a]
	b*		32±1 [a]	29±2 [ab]	29±3 [ab]	27±1[b]	29.0±2 [ab]
	L*	UV-C pre-packaging	73±1 [ab]	72±2 [ab]	74±1 [a]	71±2 [b]	71±2 [b]
	a*		-4±1 [a]	-4±0 [a]	-3±2 [a]	-3±1 [a]	-4±1 [a]
	b*		34±1 [a]	32±2 [ab]	23±1 [c]	28±2 [b]	29±1[b]
	L*	UV-C post-packaging	71±1 [a]	71±2 [a]	71±2 [a]	70±2 [a]	71±2 [a]
	a*		-4±1 [a]	-4±1 [a]	-3±1 [a]	-3±1 [a]	-3±1 [a]
	b*		33±1 [a]	31±1 [ab]	28±1 [b]	28±2 [ab]	29±2 [ab]
Preference (%)		Control	45	35	20	17	n.d.
		UV-C pre-packaging	48	50	44	48	n.d.
		UV-C post-packaging	50	50	43	52	n.d.

n.d. not determined.

[a, b, c] means in yeast loads indicated by the same letter are not significantly different ($p < 0.05$). For each colour parameter, means indicated by the same letter are not significantly different ($p < 0.05$).

It is well known that UV-C light could induce undesired surface browning in many fruits and vegetables (Gómez, Alzamora, Castro, &

Salvatori, 2010). However, no significant trend in redness and yellowness of the pineapple sticks was observed during storage in all samples. Luminosity significantly decreased in untreated samples stored for more than 10 days. A slight decrease of L* value was observed also in UV-C light treated pineapple sticks. Other authors reported significant changes in colour, associated with occurrence of translucent areas, dry surface appearance or browning even in the absence of UV-C light treatments (Marrero & Kader, 2006; Montero-Calderón, Rojas-Graü & Martín-Belloso, 2008). This result suggests that changes in the appearance of fresh-cut pineapple during storage would be attributable to the intrinsic variability of the fruits, rather than to the application of UV-C light.

Since UV-C light treatment could modify flavour of treated fruit, due to changes in volatile profile (Lamikanra & Richard, 2004), consumers were asked to evaluate the eventual presence of off-odours (data not shown). The latter were not perceived in any of the sample submitted to consumer evaluation. The high microbial load of untreated sample was expected to be correlated with typical fermented off-odour. However, pineapple flavour is also expected to increase during storage, due to fruit ripening, possibly masking eventual microbial off-odours. Consumers were also asked to express a preference between untreated samples and UV-C light treated ones, through a pair comparison preference test. Collected data are reported in Table 2. UV-C light treated samples were always preferred to the control ones and equally preferred (50% preference) to the just prepared pineapple sticks up to 3 days of storage. This result is probably due to the lower microbial load of UV-C light treated samples, allowing a longer quality retention of fresh-cut pineapple.

Based on these results, it could be hypothesized that UV-C light treatment can be successfully applied for extending the shelf-life of fresh-cut pineapple sticks well beyond the 3-5 days commonly indicated by producers of this fresh-cut fruit. The treatment could be applied even on already packaged items, avoiding any risk for post-treatment contamination.

Possible Technical Solutions for UV-C Light Implementation in Production Line

Whether the UV-C light treatment is applied on unpacked fruit or on pre-packed items, equipment, process and packaging conditions have to be carefully organized and controlled in order to guarantee a 100% exposure of fruit surface, avoiding shade effects among items, between equipment and treated fruit and between packaging materials and treated fruit. These aspects should be carefully considered when developing UV-C light equipment for fresh-cut pineapple treatment.

Unfortunately, UV-C light equipment available on the market are not specifically designed for this purpose. In fact, they are generally addressed to disinfection of air, water and beverages as well as of surfaces including packing materials and conveyor belts. However, some technological solutions are available also for disinfection of surfaces of solid food. The latter are commonly based on UV-C disinfection lamps mounted over conveyors and tunnels, filling/slicing machines, dough provers and cutting blades (*https://www.uv-light.co.uk/food/*). Some examples of foods which are UV disinfected include bread, baked goods, hard cheeses, eggs, meat, salads, nuts, seafood, fruits and vegetables (Bintsis et al., 2000). The main issue of nowadays available equipment is that the bottom side of food item is difficultly treated. This is overcome using portable lamps but this solution hardly fits with continuous food processing. Alternatively, the equipment consists of a specific tool for turning food. However, this solution is not applicable to fragile food that can be easily damaged. A possible solution could consist of a tunnel with a conveyor provided with a part in which 100% surface is irradiated. For example, the irradiation of the food bottom could be achieved through a UV-C transparent support (Figure 3a). Alternatively, the food item could be exposed to UV-C light while passing through a specific chamber equipped with UV-C lamps (Figure 3b).

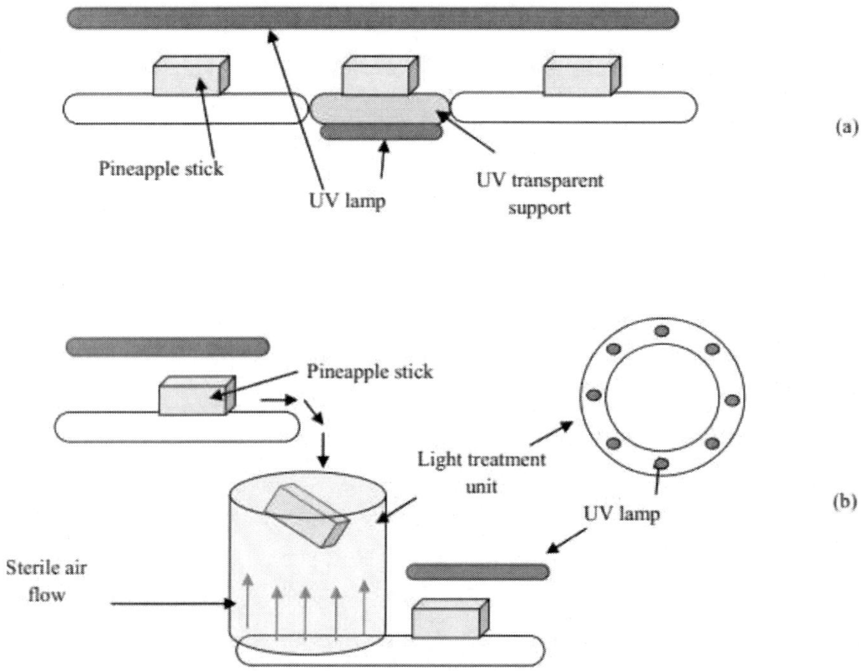

Figure 3. Possible implementation of equipment allowing a 100% UV-C light irradiation of fresh-cut pineapple stick surface. They are based on a quartz conveyor (a) or a chamber in which the pineapple stick passes through (b).

The UV-C light treatment is usually applied immediately before packaging, to maximally reduce post-treatment contaminations. To the same aim, packaging materials are UV-C light decontaminated too (*https://www.uv-light.co.uk/food/*). Nevertheless, the choice of a proper packaging material, transparent to UV-C light could allow decontamination treatment to be applied on pre-packed items, eliminating any risk for recontamination. Also in this case, precautions must be taken to maximize treatment efficacy. For instance, this solution requires packaging single fruit sticks to avoid shadowing effects among food items. In addition, it is necessary to adopt unprinted packaging materials allowing transmittance of UV-C light. Moreover, the increase in packaging thickness due to seal is expected to locally reduce material UV-C light permeability and thus a higher UV-C light fluence must be applied to compensate this screening effect. Finally, the application of a label or a secondary printable package to provide the consumer with the necessary information is required.

CONCLUSION

Fresh-cut pineapple is a highly appreciated convenient food. One of the main issues of this product is the limited shelf-life. UV-C light technology can be efficaciously used for shelf-life extension of fresh-cut pineapple. The application of low fluences can allow a substantial decrease in microbial growth during fresh-cut pineapple storage without modifying its sensory characteristics. The treatment can be advantageously applied even on pre-packed fruit, limiting the risk for post-treatment contamination and reducing fruit leakage on processing lines, making thus the process cleaner. To this aim, dedicated equipment has to be developed, allowing the entire surface of the product to be irradiated.

REFERENCES

Alothman, M., Bhat, R., & Karim, A. A. (2009). UV radiation-induced changes of antioxidant capacity of fresh-cut tropical fruits. *Innovative Food Science and Emerging Technologies,* 10(4), 512–516.

Azarakhsh, N., Osman, A., Ghazali, H. M., Tan, C. P., & Mohd Adzahan, N. (2012). Optimization of alginate and gellan-based edible coating formulations for fresh-cut pineapples. *International Food Research Journal*, 19(1), 279–285.

Bintsis, T., Litopoulou-Tzanetaki, E., & Robinson, R. K. (2000). Existing and potential applications of ultraviolet light in food industry: A critical review. *Journal of the Science of Food and Agriculture*, 80, 637–645.

Cai, Y., Shi, L., Chen, W., & Yang, Z. (2015). Effect of UV-C treatment on fruit quality and active oxygen metabolism of postharvest strawberry fruit. *Journal of Chinese Institute of Food Science and Technology*, 15(3), 128–136.

Chonhenchob, V., Chantarasomboon, Y., & Singh, S. P. (2007). Quality changes of treated fresh-cut fruits in rigid modified atmosphere packaging containers. *Packaging Technology and Science*, 20, 27–37.

Cook, R. (2015). *Trends in the marketing of fresh produce and fresh-cut products,* (Department of Agriculture and Resource Economic, University of California Davis), https://arefiles.ucdavis.edu/uploads/filer_public/2014/10/08/freshcut20140926finalnewmarycook.pdf. Accessed 24/01/2017.

Fonseca, J. M., & Rushing, J. W. (2006). Effect of Ultraviolet-C light on quality and microbial population of fresh-cut watermelon. *Postharvest Biology and Technology*, 40, 256–261.

Freitas, A., Moldão-Martins, M., Costa, H. S., Albuquerque, T. G., Valente, A., & Sanches-Silva, A. (2015). Effect of UV-C radiation on bioactive compounds of pineapple (Ananas comosus L. Merr.) by-products. *Journal of the Science of Food and Agricolture*, 95, 44–52.

Gardner, D. W. M., & Shama, G. (2000). Modelling UV-induced inactivation of microorganisms on surfaces. *Journal of Food Protection*, 63, 63–70.

Gómez, P. L., Alzamora, S. M., Castro, M. a., & Salvatori, D. M. (2010). Effect of Ultraviolet-C light dose on quality of cut-apple: microorganism, colour and compression behaviour. *Journal of Food Engineering*, 98, 60–70.

González-Aguilar, G. A., Ruiz-Cruz, S., Cruz-Valenzuela, R., Rodríguez-Félix, A., & Wang, C. Y. (2004). Physiological and quality changes of fresh cut pineapple treated with antibrowning agents. *LWT–Food Science and Technology*, 37, 369–376.

https://www.uv-light.co.uk/food/. *UV disinfection in food processing and packaging*. Accessed 24/01/2017.

International Fresh-cut Produce Association (*IFPA*). 2000. Fresh-cut facts, http://www.creativew.com/sites/ifpa/fcf.html. Accessed 24/01/2017.

Jennylynd B. J., & Ngarmsak, T. (2010). Processing of fresh-cut tropical fruits and vegetables: A technical guide. *Food and Agriculture Organization of the United Nations*.

Keyser, M., Müller, I. A., Cilliers, F. P., Nel, W., & Gouws, P. A. (2008). Ultraviolet radiation as a non-thermal treatment for the inactivation of microorganisms in fruit juice. *Innovative Food Science and Emerging Technologies*, 9, 348–354.

Lamikanra, O., & Richard, O. A. (2004). Storage and ultraviolet-induced tissue stress effects on fresh-cut pineapple. *Journal of the Science of Food and Agriculture*, 84(14), 1812–1816.

Manzocco, L., & Nicoli, M. C. (2015). Surface processing: existing and potential applications of ultraviolet light. *Critical Reviews in Food Science and Nutrition*, 55(4), 469–484.

Manzocco, L., Plazzotta, S., Maifreni, M., Calligaris, S., Anese, M., & Nicoli, M. C. (2016). Impact of UV-C light on storage quality of fresh-cut pineapple in two different packages. *LWT–Food Science and Technology*, 65, 1138–1143.

Manzocco, L., Da Pieve, S., Bartolomeoli, I., & Maifreni, M. (2011). Shelf life extension of fresh-cut fruit by UV-light exposure. University of Udine. Available at: http://www.icef11.org/content/papers/nfp/NFP114.pdf. Accessed 24/01/2017.

Marrero, A., & Kader, A. A. (2006). Optimal temperature and modified atmosphere for keeping quality of fresh-cut pineapples. *Postharvest Biology and Technology*, 39, 1639–168.

Montero-Calderón, M., Rojas-Graü, M. A., & Martín-Belloso, O. (2008). Effect of packaging conditions on quality and shelf-life of fresh-cut pineapple (Ananas comosus). *Postharvest Biology and Technology*, 50, 182–189.

Pan, J., Vicente, A. R., Martinez, G. A., Chaves, A. R., & Civello, P. M. (2004). Combined use of UV-C irradiation and heat treatment to improve postharvest life of strawberry fruit. *Journal of the Science of Food and Agriculture*, 84, 1831–1838.

Pan, Y., & Zu, H. (2012). Effect of UV-C radiation on the quality of fresh-cut pineapples. *Procedia Engineering*, 37, 113–119.

Raybaudi-Massilia, R. M., Mosqueda-Melgar, J., Soliva-Fortuny, R., & Martín- Belloso, O. (2009). Control of pathogenic and spoilage microorganisms in fresh-cut fruits and fruit juices by traditional and alternative natural antimicrobials. *Comprehensive Reviews in Food Science and Food Safety*, 8, 157–180.

Regulation EC No 258/97 of the European parliament and of the Council of 27 January 1991 concerning novel foods and novel food ingredients.

Rohrbach, K. G., & Johnson, M. W. (2003). Pests, diseases and weeds. In D. P. Bartholomew, R. E. Paull, & K. G. Rohrbach (Eds.), *The pineapple: botany, production and uses* (pp. 203–251). New York: CABI Publishing.

Sastry, S. K., Datta, A. K., & Worobo, R. W. (2000). Ultraviolet light. *Journal of Food Science*, 65, 90–92.

U.S. Food and Drug Administration. (2000). 21 CFR Part 179. *Irradiation in the production, processing and handling of food*. Federal Register, 65: 71056–71058.

Yaun, B. R., Sumner, S. S., Eifert, J. D., & Marcy, J. E. (2004). Inhibition of pathogens on fresh produce by ultraviolet energy. *International Journal of Food Microbiology*, 90, 1–8.

In: Tropical Fruits
Editors: C. Stewart Bogsan et al.
ISBN: 978-1-53612-885-7

Chapter 5

IONIZING ENERGY FOR SHELF-LIFE LONGEVITY ON PINEAPPLES' READY-TO-EAT STATUS

***Marcia Nalesso Costa Harder*[1,*] *and Valter Arthur*[2]**
[1]Technology College of Piracicaba - FATEC-Piracicaba
"Dep. Roque Trevisan"/CEETEPS, Food Technology Department,
Piracicaba, São Paulo, Brazil
[2]Nuclear Energy Center in Agriculture, Piracicaba, São Paulo, Brazil

ABSTRACT

Pineapple is a kind of tropical fruit widely consumed and appreciated worldwide. To reduce pineapple postharvest losses, several technologies are employed and the ready-to-eat technology is an innovation applied in order to achieve a long period of shelf-life for fresh fruit. There is a large consumer market which demands high quality food in a short period of time. Food radiation technology is used to promote higher quality produce. The aim of this work is to present a critical review on technological researches about the use of ionizing radiation as a form of a

* Corresponding Author Email: marcia.harder@fatec.sp.gov.br.

technology to increase shelf-life in ready-to-eat pineapple and the sensorial acceptance from consumers. The conclusion is that the use of ionizing radiation, in most cases, did not present changes in the sensory characteristics of ready-to-eat pineapples.

Keywords: pineapple minimally processed, sensorial analisys, sensory acceptance, gamma rays, x-rays, electron beam

INTRODUCTION

According to Reinhardt, Souza and Cabral (2000) [1] pineapple is a kind of tropical fruit widely consumed and appreciated by people all over the world. In Thailand, pineapple production is 2.3 million tons/year and in Brazil the production is 1.7 million tons [2]. As Brazil and Thailand are the main producers, this fruit gives high economic and social importance to these countries. The growth in the exports market of pineapple is favored by the improvement of the productive systems practiced by the use of technologies that promote the quantitative and qualitative enhancement of the production and the regularity of the supply at competitive prices in the international market.

Pineaple fruit transport is usually carried out in bulk in the main producing regions.

This procedure causes losses of 5% to 10% of the fruits to the distributors (wholesalers and retailers) either by the kneading of the fruits or by the fermentation due to lack of air circulation in the fruits arranged in the internal part of the load. This loss is very significant for producers.

In order to minimize these losses, researches have been carried out extensively with the objective of increasing the shelf-life of the fruits. At the same time, researches hope to keep the fruit quality for final consumers fresh.

Another aspect that research has taken into account is the contemporary consumer's need to find ready-to-eat foods with nutritional qualities and freshness for the longest period of time possible.

As far as it concerns Silva, Spoto, Silva (2007) [3], several researches were developed with the goal of increasing the useful shelf-life of the fruits by maintaining the quality for the consumer. Numerous technologies, ranging from the use of chemicals to heat treatment, have been studied and applied in several countries. Thus, this concern is necessary because the fresh fruits are living tissues and are subject to the proper changes of maturation. Maturation changes cannot be stopped, but can be reduced by technological procedures. The visual, sensorial and physical-chemical characteristics of the fruits are the main ones for both internal and external markets. Knowledge about the changes in these characteristics after the fruits are submitted to post-harvest treatments is of great importance in maintaining quality during transportation, storage and commercialization.

The application of ionizing energy comes from the use of clean and safe processes of food preservation. It is considered a clean, accessible, safe and efficient technology applied to the process of increasing the shelf-life of agricultural products. Its application has the purpose of eliminating parasitic insects, larvae and eggs as well as for the minimization of microbial contamination by eliminating pathogenic and deterioratintig microrganisms.

Food irradiation is classified as a peaceful use of nuclear energy and although the discovery of food irradiation dates back to the discovery of radiation at the end of the 19th century, this technology was the only in-depth studied in the mid-1960s.

One of the factors that must be considered regarding foods subjected to ionizing radiation is the sensorial characteristic. In some countries, such as Brazil, legislation states that the maximum dose of radiation to be used is dependent on maintaining the original sensory characteristics of the product.

Due to these characteristics, the aim of this work is to present a critical review on the researches developed regarding the theme of using the radiation in processing pineapples and the direct consumer.

Methods

Pineapple (*Ananas comosus* L.) is an authentic specimen of the tropical and subtropical regions, native to the coastal regions of South America. It grows in Asia, Africa and North, Central and South America. The main producers are Thailand, the Philippines, Brazil, China and India [4].

Pineapples have excellent sensorial quality attributed to the presence of sugars, organic acids and volatile compounds, which gives them a characteristic flavor and aroma. It is cultivated in several tropical and subtropical regions and appreciated all over the world [5].

According to Alves et al. (2011) [6] the excellent qualitative characteristics of the fruits reflect their socioeconomic importance [7]. It is a fruit that has a high energy value regarding its high sugar composition and nutritive value due to the presence of mineral salts (calcium, phosphorus, magnesium, potassium, sodium, copper and iodine) and vitamins, mainly ascorbic acid, thiamine, riboflavin and niacin [8]. In addition, it contains cellulosic fibers that are important for digestive functioning and bromelain proteinase that facilitates the digestion of meat [4].

For Moda et al. (2008) [9] the minimally processed products are similar to the *in natura* product in terms of nutritional and organoleptic aspects, but present modifications in their natural condition due to the application of technologies such as peeling, cutting, centrifuging and packaging [10]. With pineapple's minimum processing, it is a product with almost unchanged sensory and nutritional characteristics of great convenience for immediate consumption in small individual portions [11]. However, minimal processing may reduce fruit and vegetable shelf-life due to increased surface exposition area and injury, providing nutrients to undesirable microrganisms, as well as the release of endogenous enzymes and increase respiration rate [12, 13]. Thus, other technologies must be used in conjunction with minimum processing (cooling, modified atmosphere, ionizing radiation, use of edible films, waxes) in order to optimize the process.

Minimal processing of fruits is an alternative technology to reduce post-harvest losses and contribute to the further development of agribusiness. Firstly, this process helps to keep the product fresh and to supply it in a convenient form for the consumer, and secondly, it extends shelf-life and facilitates its distribution to consumers [14].

Food irradiation has received increasing attention because of its advantages over conventional food processing methods. Irradiation can inhibit the budding of roots and tubers, infestation of parasites and insects and reduce pathogens and microrganisms, increasing the shelf-life of fruits and vegetables *in natura* or processed. Thus, irradiation complements other conservation techniques and can be performed after packaging the product, reducing the possibility of recontamination. The irradiation process entails few chemical changes in food, which are not harmful or dangerous. The variation of the nutritional value caused by irradiation depends on several factors, such as exposure dose, type of food, its packaging and treatment conditions (such as temperature during irradiation and storage time). Thus, the treatment of food by irradiation does not physically change the appearance, shape or temperature of the products if properly controlled [15].

Food irradiation is a physical treatment process consisting of submiting the food, whether packed or in bulk, to controlled doses of ionizing radiation for sanitary, phytosanitary and/or technological purposes. The process involves the exposure of food to ionizing energy from gamma rays, x-rays or electron beams. The energy involved in irradiation is insufficient to alter the atomic nucleus of the irradiated material and since the latter does not come into contact with the radioactive source, the food does not become radioactive. As the most common sources of gamma rays for food processing, the penetrating energy of radioisotopes Co^{60} and Cs^{137} causes small molecular changes similar to those occurring in the act of cooking or freezing. Energy simply passes through the food being treated and unlike chemical treatments leaves no residue [16].

Finally, the irradiation process entails minimal chemical changes in food with no harmful or dangerous effects; hence, the World Health Organization (WHO) recommends its application and use [17].

The process of food irradiation is called cold sterilization, as long as the temperature variation of processed foods is insignificant. Products that have been irradiated can be transported, stored or even consumed immediately after treatment, because as previously described, it is a technology that leaves no residue.

It should be emphasized that the irradiation process does not improve the quality of the raw material. Therefore, it is still necessary to maintain Good Manufacturing Practices (GMPs).

Universally, for all food submitted to food irradiation, the Radura symbol can be used to identify the process (Figure 1). This should be placed on irradiated food packages in many countries. The Radura symbol is listed in the Codex Alimentarius Standard on Labelling of Prepackaged Food. The FDA requires that foods that have been irradiated have the "Radura" logo along with the statement "Treated with radiation" or "Treated by irradiation".

In the contemporary space of worldwide legislation, the most modern ones state that the minimum dose absorbed must be sufficient to achieve the intended purpose: the maximum absorbed dose should be lower than the one that would influence the functional properties and/or the sensory attributes of the food.

Owureku-Asare et al. (2014) [14] reinforce that the application of ionizing radiation for preservation of fruits and vegetables has proved effective in improving the shelf-life of fruits and reducing spoilage.

Figure 1. The Radura symbol.

Lobo and Paull (2017) [18] add that expert groups of national and international organizations as well as many regulatory agencies have generally concluded that irradiated food is safe and wholesome, and with commonly used dosing levels it does not present any enhanced toxicological, microbiological or nutritional hazards to the food beyond those brought about by conventional food processing techniques. These experts have agreed that irradiation of food for microbial safety should be carried out under Good Manufacturing Practices (GMPs) and Good Irradition Practices (GIPs) (Farrar et al., 1993), including the application of this technology to pineapples.

According to Dutcosky (1996) [19] the sensorial analysis is an evaluation of the organoleptic characteristics of foods carried out based on the human senses. The main objective is to measure the impact created by the physical properties of food, affecting consumer preference. In general, the parameters most commonly evaluated in the sensory analysis are: color, smell, appearance, texture and flavor.

According to Meilgaard, Vance and Civille (1999) [20] one of them can infer the preference of the consumers according to the relative values of acceptability obtained in the affective tests, that is, the samples with higher grades will be more preferred by the tasters.

Fellows (2006) [21] reports that volatile compounds are also produced by the action of heat, ionizing radiation, oxidation or enzymatic activity in proteins, fats and carbohydrates. However, when the sensorial evaluations were analyzed, many tasters could not identify such compounds in relation to pineapples of the control groups and groups treated with 1.5kGy and 3.0kGy. In addition, the scores attributed by the tasters in the sensorial analysis were between 6.2 and 8.4, demonstrating fruit acceptability regardless of the treatment.

In their studies, Hajare et al. (2006) [22] concluded that radiation processing with 2kGy did not affect significantly the nutritional value as well as the sensory quality of minimally processed pineapple samples.

Silva, Spoto and Silva (2007) [3] working with irradiated pineapples at doses of 100 and 150Gy at a dose rate of 175Gy/h concluded that the different doses applied had few significant influences on the sensory characteristics of pineapple; however, the results found for irradiated fruits were, in general, not favorable when observed from a quality point of view.

Martins et al. (2008) [23] obtained good acceptance for samples of irradiated pineapple up to the dose of 2.5kGy, not presenting any significant statistical difference between the control and irradiated samples.

Arvanitoyannis (2010) [24] in his work cites a very similar research to the one of Martins et al. (2008) [23], reinforcing the conclusions presented in this work.

On the other hand, Moda et al. (2008) [9] in their research of irradiating pineapples with doses of 1 and 2kGy, concluded that the control presented the best notes during the sensorial evaluation, as the better accepted sample by the tasters in the assessed items.

Fabbri (2014) [24] in his evaluation of irradiated fruits for immunosuppressed feeding observed that minimally processed pineapples responded well to the irradiation stimulations, presenting good physicochemical, microbiological and sensorial results, including positive purchase intentions in the case of commercialization. The 3.0kGy doses are seen as the best results; the production criteria is adhered to from the raw material stage to the final product stage with strict processing control using the Good Manufacturing Practices in all stages.

Studies carried out by Owureku-Asare et al. (2014) [14] confirm previous studies, since they conclude that the radiation processing of minimally processed smooth cayenne pineapples at 2kGy and 3kGy did not change its physico-chemical and sensory quality.

Based on the research presented, it can be concluded that the use of ionizing radiation in the dose of 3kGy, in most cases, did not present changes in the sensory characteristics of ready-for-eat pineapples, as they were accepted by the consumers and indicated for immunosuppressed patients.

Acknowledgments

To Prof. Mara Regina Miranda for the helpful considerations and Larissa Nalesso Costa Harder.

References

[1] Reinhardte, D. H., Souza, L. F. S., Cabral, J. R. S. 2000. *Brazilian* Fruits. Embrapa. P. 77.

[2] SEAPA. Reports. Available from: http://www.agricultura.mg.gov.br/images/Arq_Relatorios/Agricultura/2016/Mai/perfil_abacaxi_mai_2016.pdf. (Accessed 12 Feb., 2017).

[3] Silva, J. M., Spoto, M. H. F., Silva, J. P. 2007. Effect of the ionizing radiation in the sensorial characteristics of the pineapple. *Food Sci. Technol.* 27, 710-716.

[4] Figueiredo, R. M. F., Queiroz, A. J. M., Noronha, M. A. S. 2003. Armazenamento de abacaxi minimamente processado. *Rev. Bras. Prod. Agroind.* [Minimally processed pinapple storage. *Agroindustry Production Brazilian Journal]*, 15, 95-103.

[5] Andrade, A. P. S. 1999. *Padrões de identidade e qualidade para o abacaxi, a goiaba e o mamão*. M.Sc. Dissertation, Federal University of Viçosa [*Identity and quality standards for pineapple, guava and papaya.* M.Sc. Dissertation, Federal University of Viçosa. P. 124], p. 124.

[6] Alves, E. D., Maciel, L. P., Pinto, A. S. O., Franco, T. C. M., Bastos, C. T. R. M., Silva, L. H. M. 2011. Evaluation of nutritional quality and total polyphenols content from pineapple (Smooth Cayenne) as function of post-harvest storage temperature. *Rev. Bras. Pesq. Alim.*, 2, 128-134.

[7] Carvalho, V. D., Botrel, N., Gorgatti-Netto, A. 1996. *Características da fruta para exportação*. In: Abacaxi para exportação: procedimentos de colheita e pós-colheita [*Characteristics of the fruit*

for export. In: Pineapple for export: harvest and post-harvest procedures]. Embrapa-SPI, pp. 16-27.

[8] Franco, G. 2005. *Tabela de composição química dos alimentos*. 9th ed. [*Table of chemical composition of foods*. 9th ed.] Atheneu, p. 307.

[9] Moda, E. M., Pilon, L., Zocchi, S. S., Spoto, M. H. F. 2008. *Qualidade físico-química e sensorial de abacaxi minimamente processado e irradiado* [*Physico-chemical and sensorial quality of minimally processed and irradiated pineapple*]. *B. do CEPPA*, 26, 267-276.

[10] Chitarra, M. I. 2000. *Processamento mínimo de frutos e hortaliças* [*Minimal processing of fruits and vegetables]*. UFLA/FAEPE, p. 119.

[11] Bastos, M. S. R., Souza-Filho, M. S. M., Alves, R. E., Filgueira, H. A., Borges, M. F. 2000. *Processamento mínimo de melão e abacaxi*. In: Encontro de processamento mínimo de frutas e hortaliças [*Minimal processing of melon and pineapple*. In: Meeting of Minimal Processing of Fruits and Vegetables], 89-94.

[12] Rosa, O. O., Carvalho, E. P. 2000. Características microbiológicas de frutos e hortaliças minimamente processadas. *B. Soc. Bras. Ciência e Tecnol. Alim.* [Microbiological characteristics of minimally processed fruits and vegetables. *Brazilian. Society of Science and Food Technology]*, 34, 84-92.

[13] Bonnas. D. S. 2002. Qualidade do abacaxi cv. 'Smooth Cayenne' minimamente processado, embalado sob atmosfera modificada. [Quality of pineapple cv. 'Smooth Cayenne' minimally processed, packaged under modified atmosphere], PhD Thesis, Federal University of Lavras, p. 100.

[14] Owureku-Asare, M., Adu-Gyamfil, A., Agyei-Amponsah, J., Agbemavor, W. S. K., Adom-Mensah, J. B., Acquah, S., Quayson, E., Saalia, F. 2014. Effect of gamma irradiation treatment and storage on physico-chemical, microbial and sensory quality of minimally processed pineapple (*Ananas comosus*). *Br J Appl Sci Technol.*, 19, 2752-2761.

[15] Verruma-Bernardi, M. R., Spoto, M. H. F. 2003. Efeito da radiação gama sobreo perfil sensorial do suco de laranja [Effect of gamma radiation on the sensory profile of orange juice]. *Food Sci Technol.*, 23, 23-29.

[16] Araújo, L. M. 2014. Determinação dos efeitos da radiação gama nos componentes majoritários e na diâmica molecular do fruto tucumã (Astrocaryum vulgare Mart.) irradiado [Determination of the effects of gamma radiation on the major components and molecular diameters of the tucumã fruit (Astrocaryum vulgare Mart.) Irradiated]. PhD Thesis. Federal University of Rio de Janeiro, p. 126.

[17] Xavier, A. M., Lima, A. G., Vigna, C. R. M., Verbi, F. M., Bortoleto, G. G., Goraieb, K., Collins, C. H., Bueno, M. I. M. S. 2007. Marcos da história da radiotividade e tendências atuais [Landmarks of the history of radioactivity and current trends]. *Quim. Nova,* 30, 83-91.

[18] Lobo, M. G., Paull, R. E. 2017. *Handbook of pineapple technology: production, postharvest science, processing and nutrition*. Wiley, p. 672.

[19] Dutcosky, S. D. 1996. *Análise sensorial de alimentos [Sensory analysis of food]*. Champagnat, p. 239.

[20] Meilgaard, M., Vance, G., Civille, B. T. C. 1999. *Sensory evaluation techniques*. 3rd ed. Boca Raton: CRC Press, 231-255.

[21] Fellows, P. J. 2104. *Tecnologia do processamento de alimentos - princípios e prática.* [*Food processing technology - principles and practice*], 2nd ed. Artmed, p. 602.

[22] Hajare, S. N., Dhokane, V. S., Shashidhar, R., Saroj, S., Sharma, A., Bandekar, J. R. 2006. Radiation processing of minimally processed pineapple (*Ananas comosus* Merr.): effect on nutritional and sensory quality. *J. Food Sci.*, 71, 501-505.

[23] Martins, C. G., Aragon-Alegro, L. C., Behrens, J. H., Souza, K. L. O., Vizeu, D. M., Hutzler, B. W., Destro, M. T. 2008. Acceptability of minimally processed and irradiated pineaple and watermelon among Brasilian consumers. *Rad. Phys. Chem*, 77, 825-829.

[24] Arvanitoyannis, I. S. 2010. *Irradiation of Food Commodities: Techniques, Applications, Detection, Legislation, Safety and Consumer Opinion*. Academic Press/Elsevier, p. 736.

[25] Fabbri, A. D. T. 2014. A microbiológica físico-química e sensorial de salada de frutas irradiada pronta para o consumo de imunocomprometidos [The microbiological physicochemical and sensory irradiated fruit salad ready for consumption of immunocompromised]. PhD Thesis. University of São Paulo, p. 114.

In: Tropical Fruits
Editors: C. Stewart Bogsan et al.
ISBN: 978-1-53612-885-7

Chapter 6

The Effects of Gamma Radiation in the Post-Harvest Conservation of Pineapple Cv. Smooth Cayenne

Valter Arthur[1,*] and Marcia Nalesso Costa Harder[2]
[1]Nuclear Energy Center in Agriculture, Piracicaba, São Paulo, Brazil
[2]Technology College of Piracicaba—FATEC-Piracicaba "Dep. Roque Trevisan"/CEETEPS, Food Technology Department, Piracicaba, São Paulo, Brazil

Abstract

Pineapples are consumed worldwide thus post-harvest techniques still ne to be improved especially in relation to fruit transport to distribuition centers up to end-consumers. The use of ionizing energy to increase the shelf life of fruits has been extensively studied. The goal of this paper was to present a critical review survey about technological researches concerning ionizing radiation as a form of quarantine treatment of pineapple Smooth Cayenne variety. According to the results

* Corresponding Author Email: arthur@cena.usp.br.

we can conclud that it is possible to use the ionizing radiation for the purpose of post-harvest treatment in cv. Smooth Cayenne peneaple.

Keywords: ionizing energy, Hawaiian pineapple, food radiation, shelf life, good production practices

INTRODUCTION

Most fruit production is worldwide characterized by a wide variety of species being planted, in temperate climate regions with their production and consumption concentrated in the northern hemisphere. Relations in the fruit and vegetable trade in the Northern Hemisphere are complementary. Among the producing countries, Brazil is the third in the prodution ranking. However, around 40% of what is produced can be lost. The factors that may contribute to these values are inadequate use of soil and plant management techniques, lack of storage structure, logistics, inappropriate packaging and lack of knowledge of the producer (in raltion to appropriate techniques).

The three largest producers are: China, India and Brazil, together these countries account for 43.6% of world production and their productions are mainly directed to their home markets. China, in 2010 harvested 190.2 million tons which represented 26.1% of the world fruit production. Other important crops in China are those of watermelon; apple; mango; melon; tangerine; pear; peach; nectarine and plum. The second largest fruit producer, India, whose harvests of 86.0 million tonnes of fruit participated with 11.8% of total production. It stand out in this country, other crops such as Banana; Coco; Mango; Pineapple; Lemon/Limes and Cashew. Brazil which ranks the third position in the world production ranking is responsible for 5.7% of the volume harvested with a production of 41.5 million tons of fruits. Pineapple representscontributes 6.9% of the total volume of Brazilian fruit growing with 3.1 million tons [1].

The pineapple (*Ananas comosus* (L.) Merril) belongs to the Bromeliad family native from South America and specifically from the South and

Southeast of Brazil; Argentina and Uruguay [2]. It is semi-perennial plant which production cycle can vary from 14 to 24 months due to, climatic conditions however it depends on the planting time; type and weight of seedlings used and also the adopted cultural practices. Its growing process is composed, basically of two phases, one of formation of the plant or unproductive period and one of main production with only one fruit per plant. Thailand; Brazil and the Philippines are the main producers concentrating 40% of world production [3].

The study of pineapple post-harvest conservation from different production systems is necessary, to boost agribusiness, either through the possibility of supplying the consumer market or to increase the export market of fresh fruits.

Smooth Cayene is most known by European and American importers and consumers, indeed their markets are considered the most important ones. The fruits are cylindrical; yellow squash; high acidity and high sugar content; with great importance for the industry due to its physicochemical characteristics and it is known as hawaiian pineapple [4, 5, 6].

Ionizing radiation as post-harvest handling of horticultural products has been considered an important inovation. The application is carried out with the main objective of the disinfestation of grains and fruits. Recently, it has been considered as an alternative treatment to increase the useful life of some fruits and vegetables. In many countries the irradiation is effective in reducing microbial load, particularly pathogenic microorganisms, pest control and fungicide and it has been also recommended as part of programs to increase food safety as an effective method mainly in fresh fruit [7].

When fruits are submitted to these kinds of treatments using ionizing radiation mostly with aims to act as a quarantine treatment it is necessary to consider the physiological behavior of these fruits because of this treatment can cause loss of visual quality; nutritional and organoleptic of the products. Thus, it is necessary to establish the ideal dose for each cultivar which is a challenge for researchers in this area since there are several factors that interfere with the results of any post-harvest treatment. The chemical and physical effects caused by the interaction of ionizing radiation with the fruit become essential when it is desired to increase the irradiated fruit markets [4].

The aim of this work was to present a critical review survey on technological researches that used ionizing radiation as a form of quarantine treatment of pineapple variety Smooth Cayenne.

METHODS

For this work a bibliographical survey was carried out critically on studies that used ionizing radiation as a post-harvest treatment for pineapple cv smooth cayenne and it was evaluated its effects on fruits.

The commercial species of pineapple is *Ananas comosus*, a fruit of great commercial acceptance in the international market due to its organoleptic characteristics and external appearance.

Internationally the fruit of the cultivar Smooth Cayenne is the most important pineapple variety, hence it represents 70% of the fruit world production. Also known as Hawaiian pineapple, it is characterized by having a cylindrical shape; their leaves are usually needle tipped and bearing sharp upcurved spines on the margins; pale yellow flesh; high contents of sugars and acids; with citric acid being the main one. These characteristics make this cultivar suitable for both industry and *in natura* consumption. However, it has the disadvantage of being highly susceptible to cochineal and fusariosis [4].

The quality of pineapple is one of the main factors that undermine the possibility for Brazilian farmers sell the fruit abroad export, which is fundamental for its effective participation in international trade. Planting and harvesting techniques have improved greatly in recent years, but only now attention has been paid to the scientific aspect of post-harvest quality maintenance, which is directly related to transport and storage.

Among the several factors contributing to the maintenance of quality and the incidence of post-harvest losses of fruits, some aspects have to be considered, such as the initial quality of the product; the temperature at which the product was handled; stored; transported and distributed; relative humidity post-harvest environment; the use of controlled or modified atmosphere during transportation and storage; chemical treatments used for the control of physiological disorders; hot treatment to control fruit losses; packaging and handling systems.

Several studies are being carried out in order to maintain fruit quality after harvest. Among them we can mention: The aim of these studies are: types of packaging; association with application of calcium and hydrothermal treatment; use of modified atmosphere and others. All of these studies aim to maintain quality and increase the shelf life of pineapple.

In countries, where techniques of good production practices are developed, the application of methods to maintain the quality of the fruits and to reduce the damages and post-harvest losses are usual measures. However, in countries where technologies are under development, knowledge and application of techniques to maintain fruit quality are not always successful, since the solution to many fruit handling and storage problems is linked to educational and sociological factors.

Abreu and Carvalho (2017) [8] state several post-harvest driving standards such as:

- When transporting the fruits from the crop to warehouses, it is necessary to avoid high stacks to not to cause cracks; scratches or any other type of injury that could result in rotting which reduces the quality of the product;
- Harvested fruits should not be exposed to the sun light for long periods to avoid losing moisture. The location of the storage should be as close to the crop as possible or in the central region, facilitating transportation and reducing losses;
- During and after harvest, pineapple may have quality level decrease due to the penetration of pathogens through the stalk especially fungus that cause black rot or soft rot. The fruits should be disinfected with a 4,000ppm benomyl solution in order to avoid black rot that has a rapid evolution, destroying all the plant tissue, causing an acetic odor and making the fruit became soft which

leads the peel collapse easily at the slightest pressure. Temperatures of 25ºC to 32 ºC, pH 3-6 and relative humidity 90% to 100% favor the development of the fungus;

- After the peduncles have been disinfected the fruits are ready to be packed and transported to the distribution centers.

Because they are perishable, the pineapple fruits require special care after their removal from the plant due to the occurrence of several biochemical and physiological changes at that stage. These changes are characterized by a continuous process of metabolic modifications that lead to the development of important sensorial characteristics, such as sweetening. Changes in coloration and texture of the pulp culminating in their maturation and senescence [4, 9].

For these reasons, the knowledge of the post-harvest physiology of the fruits becomes of great importance when it is desired to maintain the physical-chemical; organoleptic and nutritional qualities of the fruits. In addition to the previous content, the transportation of perishable fruits, under the condition of high temperature, also contributes to fruit losses. The reduction of fruit losses due to inadequate transport and storage would contribute to the market supply in the off-season, improving distribution and creating possibilities of distribution and marketing in further places around the world [10].

However, these good production practices are not strictly adhered to and a large number of production losses are due to the post-harvest poor processes practiced.

Contrary to this assertion, many have been carried out with the aim of increasing the shelf life of fruit as well as reducing the time of quarantine treatment when exporting pineapple. Thus, the use of ionizing energy for food irradiation is an example of this action.

Food irradiation is usually done with cobalt-60 (Co^{60}) irradiators. This equipment consists of a Co^{60} source installed in a chamber which walls are shielded with cement. Foods to be irradiated (in natura or industrialized form) are placed in special containers and, through a monorail they are carried into the irradiation chamber where they receive a dose of programmed gamma radiation, for a prefixed time. It is an efficient technique in the conservation of foods, because it reduces the natural losses caused by physiological processes (budding, maturation and aging), consequently increasing the shelf life of the food. The method also eliminates or reduces the proliferation of microorganisms, parasites and pests, without causing any harm to the food. The irradiation process entails minimal chemical changes in food, none of which is harmful or dangerous, that is why the World Health Organization (WHO) recommends its application and use [11].

Since irradiation is a post-harvest process, it can not replace the pesticides used in the crops, but it can replace chemical additives used for post-harvest disinfection of fruit, such as methyl bromide which use is doomed.

Several researchers who report the advantages of this technology highlighting among them the non-formation of residues have studied the irradiation of fruits. Since the fruit does not have contact with the source of radiation and the energy received just crosses the fruit without staying stored, which guarantees the grace the dismissal of period for consumption. Another advantage mentioned is the little alteration in flavor and lower nutritional losses in the fruits, when compared to other post-harvest treatments, which makes it possible to be picked at the right time when all its natural characteristics of flavor and coloring are developed, the opposite to what happens in the heat treatment And modified atmosphere in which the fruits may undergo organoleptic changes. In addition, the fruits can be irradiated already in their final packaging which is a a practice that avoids the risk of being contaminated again after treatment [4, 12].

According to Kaferstein and Moy (1993) [13] the use of radiation when applied in low doses has been shown as a method to prolong fruit commercial life by delaying ripening and senescence processes, as well as significantly reducing rot caused by fungus and pathogenic bacteria. Therefore, the application of radiation doses in the range of radurization, that is, lower than those that cause significant changes in the organoleptic quality of the fruits become more attractive for growing fruits that are highly perishable.

However, in the use of ionizing radiations to increase post-harvest conservation of fruits and vegetables some criteria must be considered, such as, the fruit needs to show tolerance to the desired dose and that the application of radiation is as or more more than that economical and efficient than others treatments [14].

Silva (2005) [4] in a study with the use of gamma radiation in Smooth Cayenne pineapple concluded that when the doses of 100 and 150Gy (Figure 1 and 2) were used, the effects of the doses on the quality characteristics of the treated fruits were not significant for the variables firmness; total soluble solids; total titratable acidity; pH; ratio; ascorbic acid content; chlorophyll a and b content the bark; reducing sugars, sucrose and total sugars; soluble and total pectin; phenolics compounds; polyphenyl oxidase enzyme and polygalacturonase enzyme.

The storage time had more positive results in relation to these parameters than the treatments with gamma radiation and also it has not be found the presence of fungi in the external part of the fruits.

A B C

Figure 1. External appearance of pineapple after ten (A), twenty (B) and thirty days (C). Doses of zero, 100 and 150Gy, respectively.

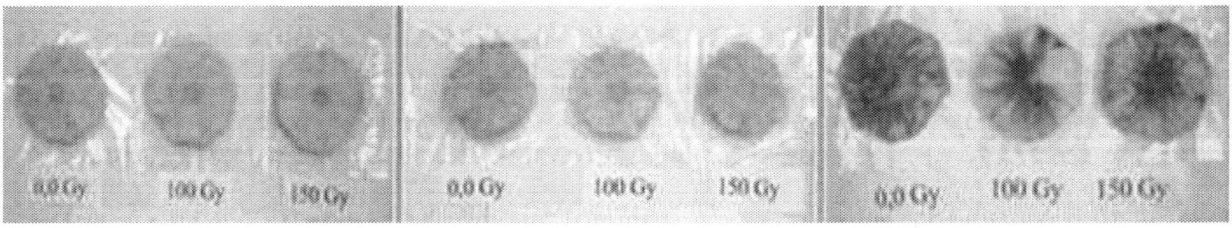

A B C

Figure 2. Internal appearance of the pineapple after ten (A), twenty (B) and thirty days (C). Doses of zero, 100 and 150Gy, respectively.

These results we confirmed by Silva, Silva and Spoto (2008) [9] research.

Stuchi-Leite, Arthur and Matraia (2005) [15] when evaluating ready-to-eat Smooth Cayenne pineapple under a dose rate of 1.084kGy/hour with doses of 0.0; 0.50; 0.75; 1.00 and 1.50kGy, evaluating the parameters of weight loss; color; firmness; soluble solids content; pH and percentage of citric acid concluded that the different doses of radiation did not significantly alter weight loss; color; texture; soluble solids content; pH and acidity of the Smooth Cayenne pineapple, under the conditions that the experiment was carried out and that doses of up to 1,50 kGy, can be used for the pineapple of the varieties Smooth Cayenne, Without prejudice to the physical and chemical characteristics of the product.

Opposed to that, Moda, Pilon, Zocchi e Spoto (2008) [16] concluded the solid soluble parameters, pH, acidity, and pH, the results obtained in the present study were similar to those of the previous study. Concerning total titrable, firmness and color, concluded that the irradiation of Smooth Cayenne pineapple ready-to-eat with 1kGy did not achieve satisfactory results for analyzes of soluble solids, titratable total acidity, pH, firmness. The control presented, in general, the best results, while the samples irradiated with 2kGy presented intermediate results in the majority of the analyzes performed.

According to the study above, can be concluded that it is possible to use ionizing radiation for the purpose of cv. Smooth Cayenne peneaple post-harvest treatment.

ACKNOWLEDGMENTS

To Prof. Mara Regina Miranda for the helpful considerations and Larissa Nalesso Costa Harder.

REFERENCES

Abreu, C. M. P., Carvalho, V. D. *Transporte e Armazenamento* [*Transport and Storage*]. Available from: http://www.ceinfo.cnpat.embrapa.br /arquivos/artigo_1538.PDF. 2017. (accessed 02 fev 2017).

Cenci, S. A. 2006. Boas Práticas de Pós-colheita de Frutas e Hortaliças na Agricultura Familiar. In: Fenelon do Nascimento Neto. (Org.). *Recomendações Básicas para a Aplicação das Boas Práticas Agropecuárias e de Fabricação na Agricultura Familiar. 1a ed.* Brasília: Embrapa Informação Tecnológica, 67-80. [Good Post-Harvest Practices of Fruits and Vegetables in Family Agriculture. In: Fenelon do Nascimento Neto. (Org.). *Basic Recommendations for the Application of Good Agricultural Practices and of Manufacture in Family Agriculture. 1st ed.* Brasília: Embrapa Information Technology, 67-80].

Chitarra, A. B.; Silva, J. M. 1990. Effect of modified atmosphere on internal browing os Smooth Cayenne pineapple. *Acta Hort.*, 485, 85-90.

Kaferstein, F. K., Moy, G. G. 1993. Public health aspects of food irradiation. *J. Public Health Policy*, 14, 149-163.

Mattiuz, B. *Fatores da pré-colheita influenciam a qualidade final dos produtos* [*Pre-harvest factors influence the final quality of the products*]. Available from : http://www.esalq.usp.br/visaoagricola/sites /default/files/va07-qualidade02.pdf. 2017. (accessed 17 fev 2017).

Melo, A. S., Aguiar-Netto, A. O., Dantas-Neto, J., Brito, M. E. B., Viégas, P. R. A., Magalhães, L. T. S., Fernandes, P.D. 2006. Desenvolvimento vegetativo, rendimento da fruta e otimização do abacaxizeiro cv. Pérola em diferentes níveis de irrigação. *Cienc. Rural* [Vegetative development, fruit yield and optimization of pineapple cv. Pearl at different levels of irrigation. *Rural Science*], 36, 93-98.

Moda, E. M., Pilon, L., Zocchi, S. S., Spoto, M. H. F. 2008. Qualidade físico-química e sensorial de abacaxi minimamente processado e irradiado. *B. do CEPPA*, 26, 267-276. [Physico-chemical and sensorial quality of minimally processed and irradiated pineapple. *CEPPA Brazilian*, 26, 267-276]

Mostafavi, H. A., Mirmajlessi, S. M., Fathollahi, H. 2012. The potential of food irradiation: benefits an limitations. Avaible from : http://cdn.intechopen. com/pdfs/35125/InTech-The_potential_of_food _irradiation_benefits_and_ limitations.pdf. (accessed 12 fev 2017).

Ponciano, N. J., Constantino, C. O. R., Souza, P. M., Detmann, E. 2006. Avaliação econômica da produção de abacaxi (Ananas comosus L.) cultivar Pérola na região Norte Fluminense. *Rev. Caatinga*, 19, 82-91. [Economic evaluation of the production of pineapple (*Ananas comosus* L.) Pérola cultivar in the North Fluminense region. *Caatinga Journal*, 19, 82-91].

Pontes Thé, P. M., Carvalho, V. D., Abreu, C. M. P., Nunes, R. P., Pinto, N. A. V. D. 2001. Modificações na atividade enzimática em abacaxi Smooth Cayenne em função da temperatura de armazenamento e do estádio de maturação. *Ciênc. Agrotéc.*, 25, 364-370. [Modifications in the enzymatic activity in Pineapple Smooth Cayenne as a function of storage temperature and maturation stage. *Agrotechnical Science*, 25, 364-370].

Roberts, P. B. 2014. Food Irradiation is safe: half a century of studies. *Radiat. Phis. Chem.*, 105, 78-82.

SEAPA. Reports. Available from: http://www.agricultura.mg.gov.br /images/Arq_Relatorios/Agricultura/2016/Mai/perfil_abacaxi_mai_20 16.pdf. (accessed 02 fev 2017).

Silva, J. M. Effects of gamma radiation (60Co) on post-harvest storage of the pineaple Cv. Smooth Cayenne, PhD Thesis, São Paulo University, 2005, p. 130.

Silva, J. M., Silva, J. P., Spoto, M. H. F. 2008. Physico-chemical characteristics of pineapple submitted to ionizing radiation technology as a method of post-harvest conservation. *Food Sci. Technol.*, 28, 139-145.

Stuchi-Leite, D.T., Arthur, V., Matraia, C. Estudo de diferentes doses de radiação gama nas características físico-químicas de abacaxi Smooth Cayenne e Pérola minimamente processado. [*Study of different doses of gamma radiation in the physico-chemical characteristics of pineapple Smooth Cayenne and Pearl minimally processed*]. Available from: https://www.ipen.br/biblioteca/cd/inac/ 2005/full/501.pdf. (accessed 10 fev 2017).

Xavier, A. M., Lima, A. G., Vigna, C. R. M., Verbi, F. M., Bortoleto, G. G., Goraieb, K., Collins, C. H., Bueno, M. I. M. S. 2007. Marcos da história da radiotividade e tendências atuais. *Quim. Nova,* 30, 83-91. [Landmarks of the history of radioactivity and current trends. *New Chemistry Journal*, 30, 83-91].

In: Tropical Fruits
Editors: C. Stewart Bogsan et al.
ISBN: 978-1-53612-885-7

Chapter 7

Processing Pineapple Juice Using Ultraviolet Irradiation

Nor Hasni Hambali[1], Pei Chen Koh[1], Mohd Adzahan Noranizan[1,*] and Rosnah Shamsudin[2]

[1]Department of Food Technology, Faculty of Food Science and Technology, Universiti Putra Malaysia, Selangor Darul Ehsan, Malaysia

[2]Department of Process and Food Engineering, Faculty of Engineering, Universiti Putra Malaysia, Selangor Darul Ehsan, Malaysia

Abstract

Consumer preference has shifted from carbonated drinks to fruit juices. There are a variety of fruit juices currently available in the market such as apple juice, carrot juice, grape juice, mango juice, orange juice, pineapple juice and mixed fruit juices. Pineapple juice is the preferred fruit juices for its pleasing aroma and flavor. The phytochemicals in pineapple juice not only reduce the risk of oxidative damage associated with the presence of free radicals but also the risk of contracting cardiovascular, cancer, and neurological diseases. Unpasteurized fruit

* Corresponding Author Email: noraadzahan@upm.edu.my.

juices are susceptible to spoilage due to its high sugar content, which serves as substrate for microbial growth. Thermal pasteurization is a conventional food preservation method that has been widely applied on fruit juices to ensure safety thus extend the shelf life. However, conventional thermal processes can affect overall quality of the juice by changing its nutritional and biochemical properties. Ultraviolet (UV) irradiation is a non-thermal disinfection technology that is applied at a low temperature, which can potentially be used as an alternative to thermal pasteurization in the juice industry. This chapter serves as a source related to development of UV irradiation technology for pasteurization and shelf-life extension of pineapple juice to successfully obtain a final product with minimal changes of its nutritional component without neglecting the microbial safety. It discusses the design of UV reactor and Dean Effect in UV technology. This chapter also covers the aspect of microbiological and chemical safety, quality, and sensory characteristics in pineapple juice processed with UV irradiation as well as its market potential.

INTRODUCTION

Pineapple juice or pineapple juice blend is usually popular in tropical continents such as Southeast Asia and South America as it helps with quenching thirst. The juice is also popular in countries with temperate climates such as United States, Belgium, Germany, Netherlands, and Italy. Besides the refreshing taste of pineapple juice, the well-known nutrients and antioxidants seem to be an attraction especially for urban consumers and gym or health enthusiasts. Unfortunately, the shelf life of fresh pineapple juice is restricted by the enzyme and microbial activities. This can be solved through pasteurization as a preservation treatment. Thermal pasteurization is an option for treatment of most juices, but it results in various nutrient and taste degrading effects. The beneficial compounds in pineapple juice such as ascorbic acid, minerals, bromelain, and antioxidants are sensitive to heat. Thus, the use of alternative preservation processes, which do not produce heat is necessary. Among the common processes available for juice treatment are ultraviolet (UV) irradiation, high pressure processing and pulsed electric field processing.

Interest in UV pasteurization for juice processing has increased consistently in recent years because of growing consumer demand for premium quality food products. This technology has been approved by the Food and Drug Administration (FDA) for juice processing and is mentioned in the FDA Code of Federal Regulations (21 CFR Part 179.39). UV was originally used as a mean for water treatment and its application in juice processing is relatively new. Attempts on UV pasteurization of pineapple juice have been made and the effects of UV on juice quality have been evaluated.

This chapter will discuss the principles and mechanisms related to UV technology as well as its impact on sensory and storage quality of pineapple juice. The cost implication and marketing strategy are also being discussed.

MARKET OF PINEAPPLE JUICE

Total global production of pineapple is about 16 million tonnes with 12 countries account for 80% of total pineapple production. Three-quarter of the 7.4 million tonnes of pineapples traded in the world is canned pineapple and pineapple juice, which has increased four-fold from 1.3 to 5.6 million tonnes (fresh fruit equivalent) in the world trade market. Two third of pineapple export consists of single or concentrated juice and the rest is canned goods (rings or pieces). Asia countries such as Thailand, Philippines, Indonesia, and Vietnam dominate the market for pineapple (Loeillet, 2006). For pineapple juice, Thailand, Philippines, and Kenya are the largest producers of pineapple juice or pineapple concentrate (Korbech-Olesen, 1997). About 82% of pineapple juice is accounted for Thailand and Philippines in which most of the juice is exported to Western Europe and the United States. Malaysia ranks number 7 in Asia after Thailand, Philippines, Indonesia, Vietnam, China, and Taiwan for the production of pineapple juice in the world (Mansor, 2014). The high demand for pineapple juice in local and overseas suggests the need of establishing alternative preservation method in extending the shelf life pineapple juice.

PRINCIPLES AND TECHNOLOGY IN UV IRRADIATION

Basic Principle of UV Irradiation

Electromagnetic energy travels in the form of waves and consists of a broad spectrum from very long radio waves to very short gamma waves (Butcher, 2016). Figure 1 shows the various forms of energy in the electromagnetic radiation spectrum including UV light.

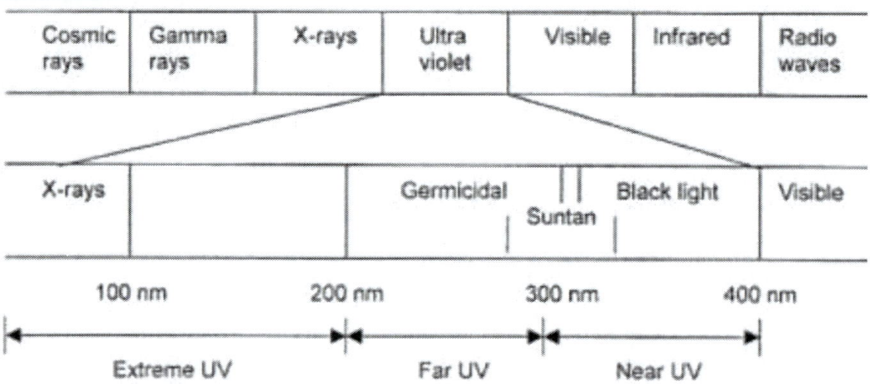

Figure 1. Division of light in electromagnetic radiation spectrum. Adapted from Guerrero-Beltrán & Barbosa-Cánovas (2004).

UV light is part of the electromagnetic spectrum with wavelengths ranging from 100 to 400 nm, which consist of UV-A (315-400 nm), UV-B (280-315 nm), UV-C (200-280 nm), and vacuum UV range (100-200 nm). Different parts of the UV spectrum have different effects, i.e., UV-A is usually related to tanning of human skin, UV-B leads to skin burning and thus skin cancer, UV-C is the germicidal portion, which inactivates microorganisms, and vacuum UV range only can be transmitted in vacuum because of its high absorption by almost all substances (FDA, 2015).

Mechanisms of UV-Light Generation

Emission of UV light is due to the transition of electrons among the energy levels in atoms or ions. An atom or ion only absorbs a discrete amount of energy. Energy is transferred to the atoms from the light due to the collisions between the free electrons and atoms. When a sufficiently high voltage is applied, the electrons in the atoms of target materials move quickly due to high kinetic energies. Upon absorption of energy by atom or ion, the electrons jump to a higher energy level. When the electrons return to lower energy level, emission of energy leads to the generation of photon (Al-Azzawi, 2006). The radiation of UV and visible light spectral range and other less-energetic radiation are known as nonionizing radiation (Koutchma et al., 2009). Figure 2 shows the absorption and emission of energy during photon generation.

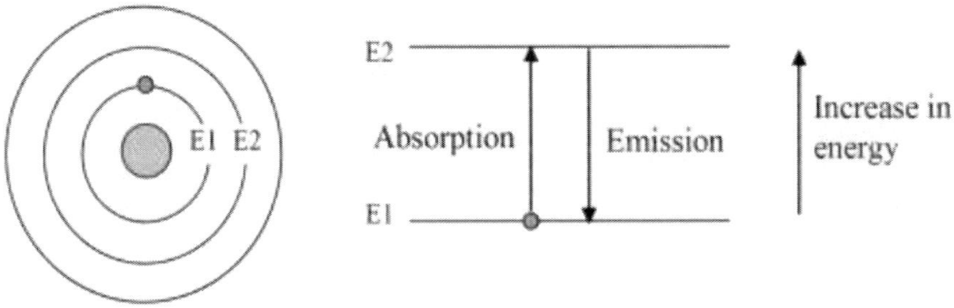

Figure 2. Absorption and emission of energy during photon generation.

The energy of the photon can be calculated based on Planck relationship.

$$E = hv$$

where h is the Planck constant, = 6.63×10^{-34} J·s and v is the frequency of light spectrum.

The energy level of an atom or ion can be varied depends on the number of electrons, protons, and neutrons present in the atom or ion and their interaction with the outer force fields. In this context, every element generates light with a unique spectrum. When the difference between the energy levels is suitable, UV is emitted as the light (Koutchma et al., 2009).

Source of UV

The commonly used UV lamps are low pressure (LP), low pressure, high output (LPHO), and medium pressure (MP) lamps. Figure 3 illustrates the schematics of these lamps. These categories of lamps are named based on the vapor pressure of mercury during the operation of lamps. The lamps are made up of a tube of vitreous silica glass with both ends sealed as the UV-transmitting envelope. The internal part of envelope contains mercury and inert gas. Alternating current is usually utilized to power the lamp. The end of each envelope has an electrode that is connected to the external via a seal (Koutchma et al., 2009).

Mercury lamps contain a low amount of mercury and an inert gas such as argon. Mercury is used due to its highest volatility among all the metal elements whereby its activation in the gas phase can be achieved at temperatures compatible with the structure of lamps. Argon is the most common filler gas with ionization energy of 15.8 eV and the lowest activated metastable state is at 11.6 eV. The collision of argon atoms and mercury atoms causes ionization of mercury and subsequently leads to light emission. The argon gases activate the starting of the discharge and stimulate the starting activation-ionization of the mercury. The mercury lamps can work at low or medium pressures (Koutchma et al., 2009). For food processing, the FDA requires the radiation sources made up of low pressure mercury lamps with 90% of the emission at a wavelength of 253.7 nm (21CFR179.39). Table 1 shows the characteristics and advantages of typical mercury lamps.

Table 1. Characteristics and advantages of typical mercury lamps. Adapted from EPA (2006)

Parameter	Type of Lamp		
	Low-pressure	Low-pressure High-output	Medium-pressure
Germicidal UV light	Monochromatic at 254 nm	Monochromatic at 254 nm	Polychromatic, including germicidal range (200-300 nm)
Mercury vapour pressure (Pa)	Approximately 0.93 (1.35×10^{-4} psi)	0.18-1.6 (2.6×10^{-5}-2.3×10^{-4} psi)	40,000-4,000,000 (5.80-580 psi)
Operating temperature (°C)	Approximately 40	60-100	600-900
Electrical input [watts per centimeter (W/cm)]	0.5	1.5-10	50-250
Germicidal UV output (W/cm)	0.2	0.5-3.5	5-30
Electrical to germicidal UV conversion efficiency (%)	35-38	30-35	10-20
Arc length (cm)	10-150	10-150	5-120
Relative number of lamps needed for a given dose	High	Intermediate	Low
Lifetime [hour (hr)]	8,000-10,000	8,000-12,000	4,000-8,000
Advantages	-Higher germicidal efficiency nearly all output at 254 nm		-Higher power output
	-Smaller power draw per lamp (less reduction in dose if lamp fails)		-Fewer lamps for a given application
	-Longer lamp life		

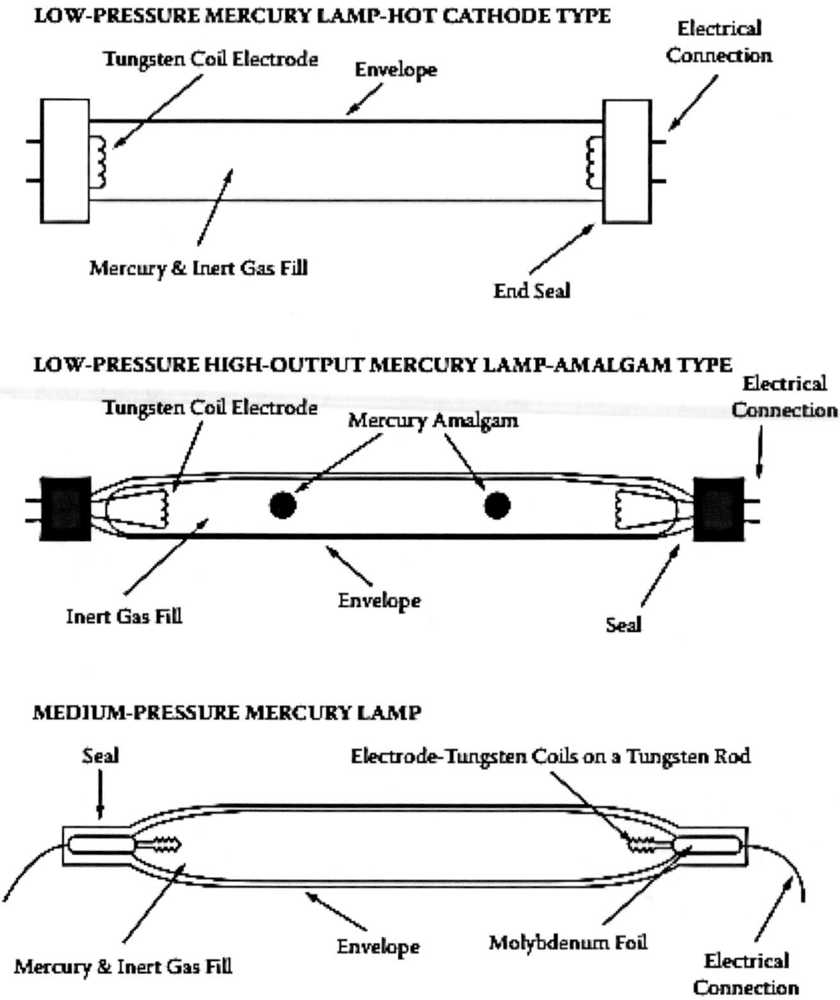

Figure 3. Schematics of LPM, LPOH, and MPM lamps. Adapted from Koutchma et al. (2009).

UV Equipment Designs for UV Treatment of Juice

UV irradiation can be applied in continuous or pulsed mode. Some UV reactor designs are for surface treatment of solid food while others are for liquid food. The type of UV technology discussed in this chapter will be continuous UV irradiation for liquid food. The design of UV reactors is based on the aim of delivering dose efficiently to inactivate microorganisms. Commercial UV reactors are made up of open or closed-channel vessels, UV lamps, lamp sleeves, UV sensors, and temperature sensors. Lamp sleeves cover the UV lamps in order to protect and insulate the lamps. Some reactors contain automatic cleaning function to remove deposits from the lamp sleeves. The dose delivered by the reactor is measured using UV sensors, flow meters, and/or UVT analyzers (EPA, 2006). A pump is necessary for controlling the flow rate of juice. In certain units, an automated self-adjustable flow rate algorithm is being incorporated into the system to ensure sufficient killing of microorganisms. Figure 4 shows the example of UV equipment. Figure 5 displays the schematics of UV irradiation liquid system.

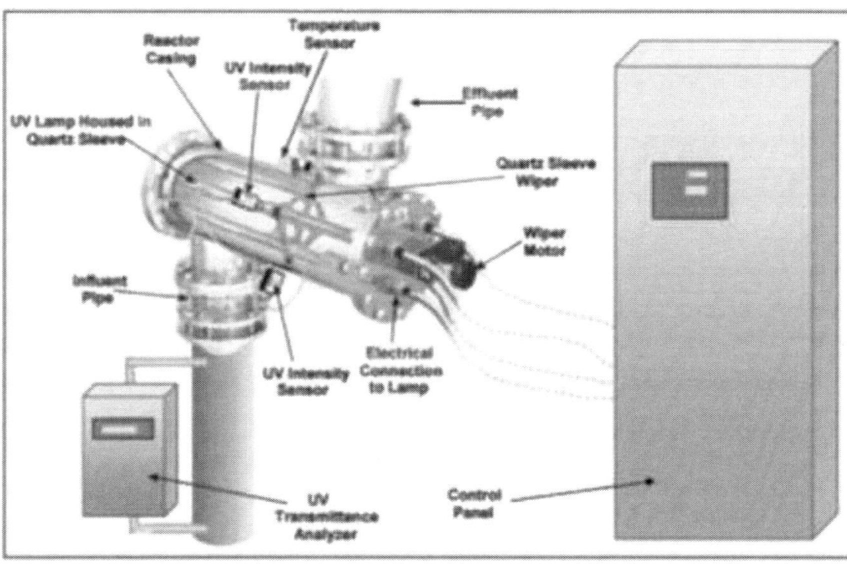

Figure 4. Example of UV equipment. Adapted from EPA (2006).

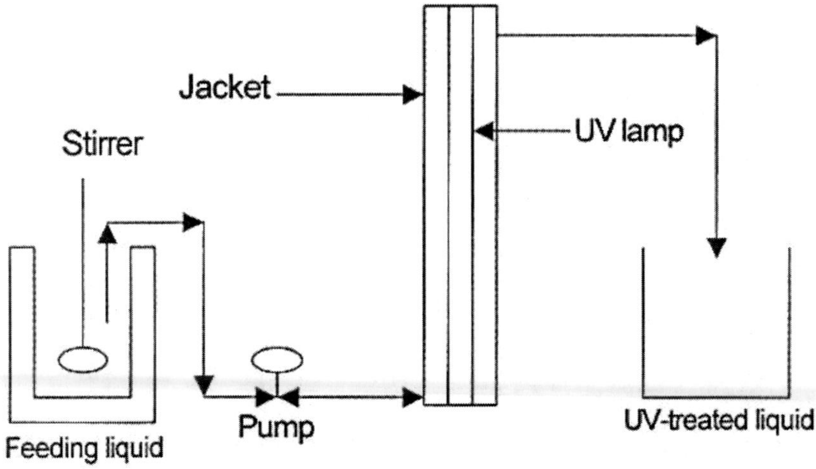

Figure 5. Schematics of UV irradiation liquid system. Adapted from Guerrero-Beltrán & Barbosa-Cánovas (2004).

For fluids with low or close to zero UV light transmittance, such as fresh fruit or vegetable juices, the acceptance of UV photons is based on the process efficiency which can be affected by design of a continuous UV system, UV dose, and its mixing efficiency. Most of the reported studies were carried out using lab-scale UV devices that are custom made based on the continuous or batch system. A well-defined UV irradiation system should contain details such as lamp characteristics, lamp power (W), lamp wavelength (nm), information of the reactor (continuous flow conditions or batch apparatus), the number of lamps used, and the number of passes through the reactor for a continuous system. For UV treatment of juices, the appropriate UV reactor design is important to avoid interference caused by high UV absorbance materials in the juice and viscosity in order to improve inactivation efficiency. Flow meters and UV sensors are crucial for dose measurement. The total delivered UV dose can be varied depending on the flow pattern of liquid. Three types of lab-scale UV continuous systems have been used, i.e., annular systems in laminar and turbulent flow, coiled-tube system in Dean flow, and Taylor-Coutte systems. The utilization of three commercial UV units, such as *CiderSure* (thinfilm; FPE, N.Y., U.S.A.), Salcor module (Dean flow in coiled tube, turbulent; Calif., U.S.A.), and SurePure (thin film, turbulent; SupePure Inc. Zug, Switzerland) are discussed in this chapter. In a thin film UV reactor, the path length for UV light to travel is reduced and therefore improves the penetration of light into samples. Thin film reactors operate with a parabolic velocity profile. Figure 6 illustrates the control panel and sensor placement of *CiderSure* 3500 UV pasteurizer. Figure 7 shows the schematics of a thin film reactor (*CiderSure*) while Figure 8 shows the schematics for laminar Taylor – Couette UV reactor. For the *CiderSure* reactor, the juice is pumped in from reservoir to UV equipment via a 0.08 cm annular gap between the inner surface of the chamber and outer surface of the quartz sleeve (Koutchma, 2008). The application of UV irradiation using *CiderSure* 3500 UV pasteurizer was effective to reduce microbial counts in various fruit juices including pineapple juice (Shah et al., 2016a; Shamsuddin et al., 2013; Sew et al., 2012; Hanisah, 2009).

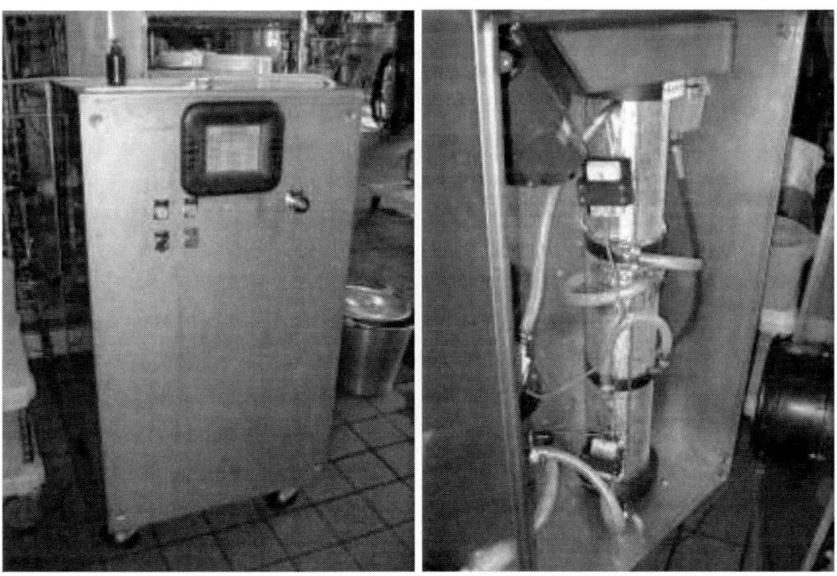

a b

Figure 6. *CiderSure* 3500 UV pasteurizer (a) control panel; (b) sensor placement.

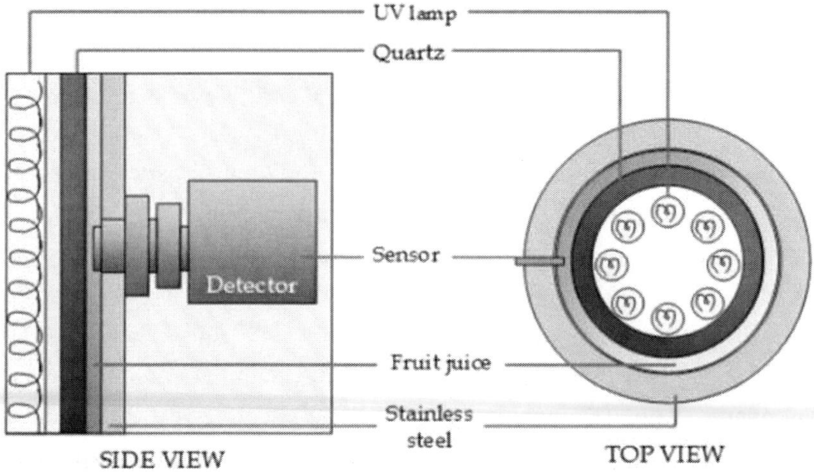

Figure 7. Schematics of *CiderSure*3500 UV pasteurizer. Adapted from Mohd Adzahan (2006).

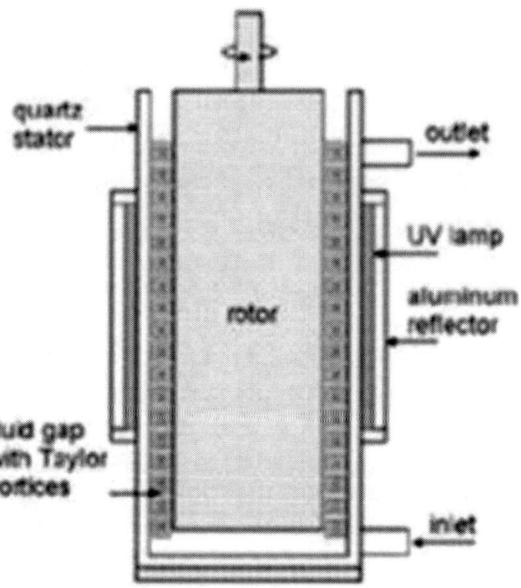

Figure 8. Schematics of laminar Taylor- Couette UV reactor. Adapted from Koutchma (2008).

Another reactor design is based on the aim of increasing turbulence in a UV reactor to allow all materials travel closer to the UV light source during treatment. The high flow rate during turbulence causes homogeneity of the flow and all parts of the juice will be treated with UV light as a result of better mixing. For turbulent channel reactor, the liquid is delivered to the UV equipment via a stainless steel chamber with UV emitting low-pressure arc-tubes mounted. Each single arc-tube is installed in a quartz sleeve in order for the liquid to flow through the sleeve in all directions. For Dean flow reactor, it consists of a coiled Teflon tube with 24 UV lamps and reflectors. The coiled tube increases turbulence and leads to secondary eddy flow, called Dean effect (Koutchma, 2008). This process will create a mixing effect and result in extra residence time for the fruit juice to be pasteurized (Muller et al., 2011). Figure 9 shows the schematic of UV pasteurization system for liquid food using the Dean Vortex principle. Mansor et al. (2014) reported that the use of Dean flow UV reactor enclosed with quartz glass was effective to prolong shelf life and maintain physicochemical quality of pineapple juice. Figure 10 shows the schematics of turbulent channel reactor and Dean flow reactor.

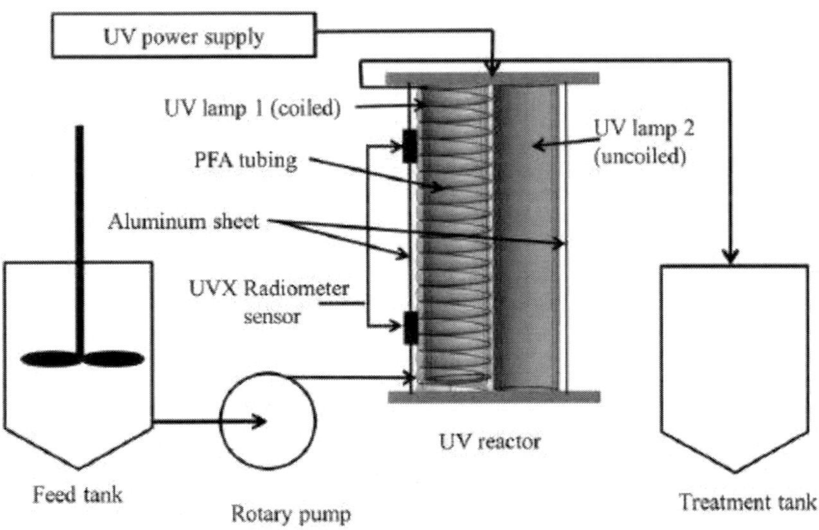

Figure 9. Schematic of UV pasteurization system for liquid food using the Dean Vortex principle. Adapted from Mansor et al. (2015) and Mansor et al. (2014).

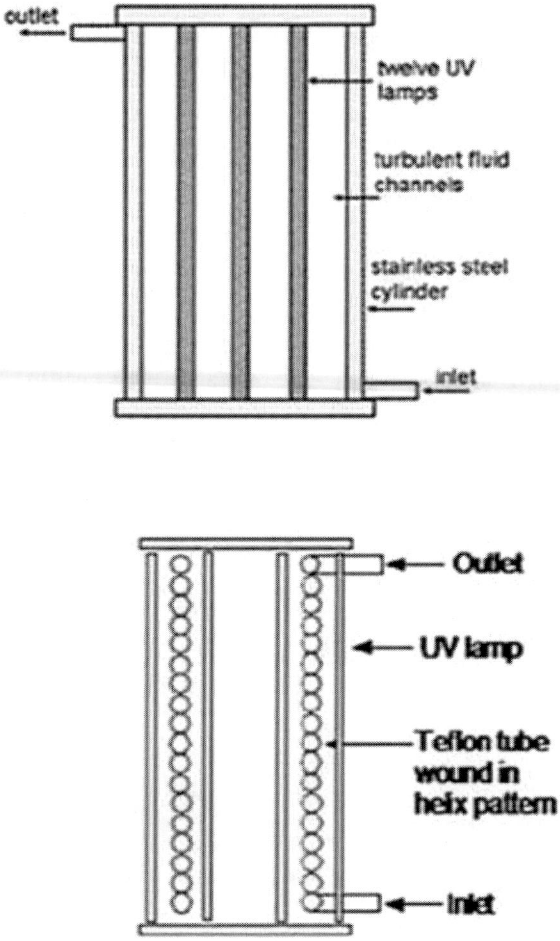

Figure 10. Schematics of turbulent channel reactor and Dean flow reactor. Adapted from Koutchma (2008).

In any design of UV reactor, quartz glass sleeve is important for proper UV processing of juices. Quartz glass or fused quartz is commonly used in thermophysical measurements for calibration and verification of thermal equipment (Sergeev et al. 1982). Wavelength lower than 350 nm can pass through the quartz glass (Mansor et al., 2015). Thus, radiation of UV-C can transmit through the quartz glass. A quartz glass has characteristics such as high purity, high thermal resistance, high resistivity, and high dielectric strength. Keyser et al. (2008) mentioned that the UV treatment of juices should utilize UV system with the UV lamp covered with a quartz glass sleeve. With this, direct contact between the surface of the lamp and the treated juice can be prevented. Quartz glass is used to minimize direct heat exposure from UV lamp to the juice. Figure 11 shows the cross section of coiled UV lamp enclosed by a quartz glass sleeve. Mansor et al. (2015) has studied the performance of UV pasteurization with and without quartz glass sleeve on physicochemical properties and microbial activity of pineapple juice. Figure 12 displays the arrangement of UV lamp for determination of middle lamp intensity in UV pasteurizer without quartz glass and UV pasteurizer with quartz glass used in the study conducted by Mansor et al. (2015). Mansor et al. (2015) found that the quartz glass was able to reduce some heat and UV radiation from the UV lamp (Table 2). It is in agreement with Langer et al. (1996) who concluded that the application of quartz glass reduced the UV portion of radiation by coating the lamp. According to Mansor et al. (2015), the UV reactor enclosed by a quartz glass sleeve was more effective in maintaining the qualities of pineapple juice in terms of color (L^* value, chroma, and hue angle), ascorbic acid content, and total soluble solids.

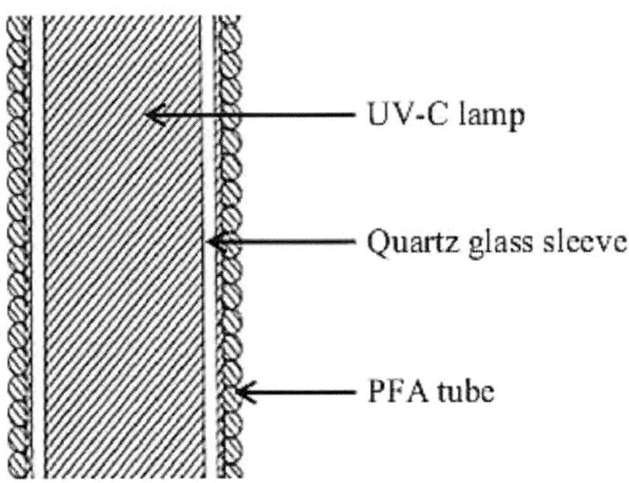

Figure 11. Cross section of coiled UV lamp enclosed by a quartz glass sleeve. Adapted from Mansor et al. (2015).

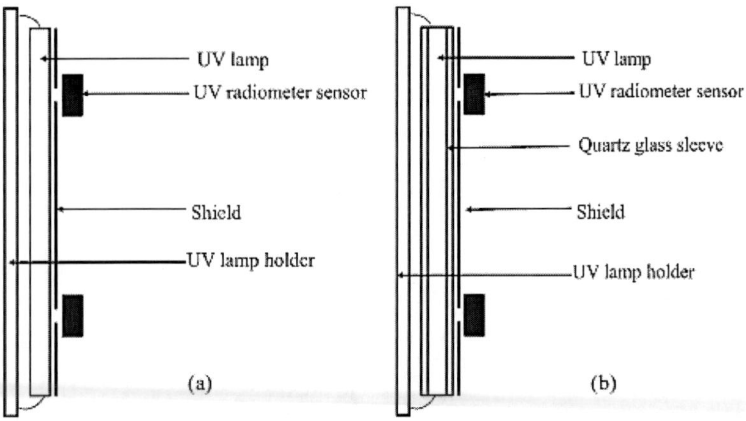

Figure 12. Arrangement of UV lamp for determination of middle lamp intensity for (a) UV pasteurizer without quartz glass and (b) UV pasteurizer with quartz glass. Adapted from Mansor et al. (2015).

Table 2. Effect of heat on the pineapple juice before and after UV treatment. Adapted from Mansor et al. (2015)

UV reactor	Without quartz	With quartz
Temperature of juice before treatment (°C)	28.6 ± 0.58	28.6 ± 0.57
Temperature of juice after treatment (°C)	62.3 ± 0.57	45.7 ± 0.58
Percentage of temperature increased (%)	117.83	59.79

Technical Aspect of UV Radiation

There are several technical aspects of UV irradiation which must be understood to ensure effective use of the technology. These aspects are discussed below.

Interaction between the Light and Materials

During emission of UV light, the UV light propagates and interacts with surrounding materials via absorption, reflection, refraction, and scattering. In UV reactor, UV light generated from the lamps will interact with parts of the reaction such as lamp, lamp sleeve, and walls and liquid to be treated. The interaction between the light and materials determines the intensity and wavelength of the UV light delivered to bacteria or chemical components in the liquid (Koutchma et al., 2009). Figure 13 displays the phenomena when light interacts with materials.

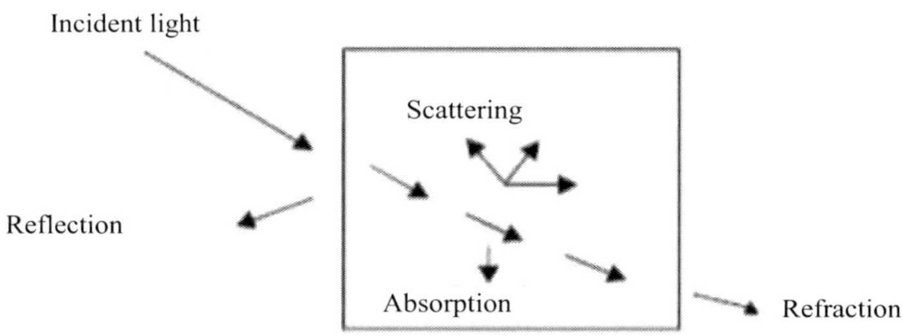

Figure 13. Phenomena when light interacts with materials. Adapted from Jawad et al. (2011).

When a light is absorbed by materials, the energy of the light is transformed into other forms of energy. In this case, there is no light available for microbial inactivation. Other interactions such as reflection, scattering, and refraction alter the UV light direction, but the light still has decontamination effect. Reflection of light occurs when the light is deflected from the surface, which can be categorized as specular or diffuse. Specular reflection happens when the light strikes on a smooth surface while diffuse reflection occurs on a rough surface with scattering of light in all directions (Kipphan, 2001). Figure 14 shows the specular and diffuse reflections when light interacts with materials. Scattering is a phenomenon involving deflection of radiation from a straight path. When the electromagnetic wave reaches an obstacle, such as electron, atom, molecule, solid or liquid particle, the electric field of the incident wave causes the electric charges in the object to be in oscillatory motion. The accelerated electric charges radiate electromagnetic energy in all directions, which lead to scattering of light (Bohren & Huffman, 2004). UV light that scattered from particles has microbial inactivation effect and thus makes light scattering to be an important phenomenon during decontamination of food liquids containing particles (Koutchma et al., 2009). For example, pineapple juice containing particles is potential to be treated with UV irradiation for shelf life extension purpose. Refraction is the phenomenon where the light changes direction when it travels from one medium to another (Figure 13). In UV treatment, refraction happens when light travels from air to quartz or from quartz to liquid (Koutchma et al., 2009).

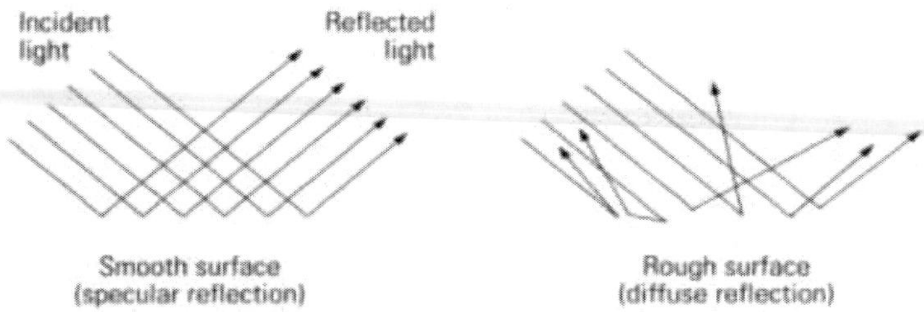

Figure 14. Specular and diffuse reflections when light interacts with materials. Adapted from Kipphan (2001).

Limiting Factors

Due to the light phenomena, the intensity of UV light decreases as the light travels leading to lower light penetration. The penetration of light into the target is reduced when the distance between the light source and target is increased. This is especially important for liquid with low transmissivity (high turbidity). Liquid with high transmissivity can be treated with UV-C radiation easily. For example, water has the highest transmissivity. On the other hand, liquid with low transmissivity due to presence of particulute materials or organic compounds poses difficulties to be treated by UV-C radiation. Small particles suspended in liquid decrease UV penetration and thus the germicidal effect of UV. UV-C light penetrates into juice at distance of about 1 mm with the light absorption at 90% (Guerrero-Beltrán& Barbosa-Cánovas, 2004).

Since light travels in a straight and parallel line, shadowing effect is another limitation of applying UV irradiation in fruit juice. In this context, the light phenomena limit UV irradiation performance in which the presence of solids or particles in the liquid will tend to block the UV light to penetrate into the liquid. The particles present at the outermost surface of the liquid will shadow and provide a protective effect to the internal part of the liquid and thus reduce penetration of UV light. As a result, the germicidal effect from UV will be affected. The shadowing effect increases with the increase of the amount of particles in a liquid (higher turbidity).

Other than that, UV light is highly absorbed by compounds containing conjugated bonds, such as aromatic ring and double ring compounds as well as compounds containing disulfide bonds. Among the examples of these compounds are vitamin A, B2, B12, C, D, E, K, carotenes, folic acid, tryptophan, and unsaturated fatty acids. In addition, sugars such as fructose, sucrose, and glucose have a high absorbance in UV range of 240-360 nm (Shah et al., 2016a). These compounds are of high concentrations in fruit juice. As such, the UV light absorbed by these compounds during the treatment of fruit juice will reduce the penetration of the light into internal part of the juice. Halim et al. (2012) attempted to pasteurize pitaya juice using *CiderSure*. Unfortunately, they failed to achieve a 5 $\log_{10}$ reduction even after using the hurdle concept. This is due to the strong color of pitaya and various UV-absorbance compounds in the juice, which limit UV penetration.

Due to all the limitations mentioned, juices passing through the UV reactors should be made into a thin film so that the path length is shorter for easier penetration. Reactors can also be designed to create turbulence flow in order for the juice to flow through UV system for disinfection. According to Muller et al. (2011), the UV-C absorption by fruit juice particles or dissolved substances can be reduced by applying Dean vortex technology, which improves the mixing effect inside the tube. The Dean vortex technology has the potential to inactivate spoilage microorganisms in cloudy fruit juice. Muller et al. (2011) proved that *L. plantarum* population was reduced in orange juice with an increase of Dean Number. Mansor et al. (2014) also found that treatment using Dean flow UV reactor was effective to prolong the shelf life of pineapple juice while maintaining its physicochemical quality.

Absorption Coefficient

The efficacy of UV irradiation on fruit juices is affected by the absorption coefficient of the juice, which influences the UV penetration and thus microbial inactivation. Juice with low UV absorption will have low absorption coefficient. The absorption coefficient is affected by the characteristics of juice, such as color and cloudiness (Murakami, Jackson, Madsen, & Schickedanz, 2006). Sew et al. (2012) mentioned that the clarified pineapple juice (16.7 cm^{-1}) had a lower absorption coefficient than pineapple juice without clarification (29.9 cm^{-1}). These findings suggest that UV light will penetrate better in clarified juice compared to non-clarified juice. Table 3 shows the absorption coefficient of pineapple juice and other beverages at 254 nm.

Table 3. Absorption coefficient (α) of pineapple juice and other beverages at 254 nm. Adapted from Sew et al. (2014)

Juice	α, cm−1	References
Distilled water	0.007-0.01	Shama (1999)
Drinking water	0.02-0.1	Shama (1999)
Apple juice	13.9	Pataro et al. (2011)
Pineapple juice (clarified)	16.7	Sew et al. (2012)
Pineapple juice (non-clarified)	29.9	Chia (2011)
Cider	30	Koutchma & Parisi (2004)
Naturally cloudy apple juice	48.4	Müller et al. (2011)
Orange juice	52.4	Pataro et al. (2011)
Blood orange juice	194.3	Müller et al. (2011)
Milk	300.0	Shama (1999)
Elderberry nectar	329.3	Müller et al. (2011)

Dosage Measurement

Since UV dose indicates the amount of UV light exposed to a liquid, measurement of the intensity of UV light is vital to ensure efficiency of

UV treatment. Measurement of UV dose is important to obtain the desired dose for each sample, to obtain data for approval by regulatory agencies, and to establish quality control procedures for the treatment plant. The measurement of UV irradiance is commonly based on energy incident per unit area normal to the beam. The UV irradiance is termed energy fluence and fluence rate. Fluence rate or irradiance is defined as the total radiant power incident from all directions exposed to a small sphere of cross-sectional area (Koutchma et al., 2009). In this context, the UV-C dose emitted from a lamp is usually expressed in the unit of W/m^2 (Guerrero-Beltrán & Barbosa-Cánovas, 2004). Meanwhile, fluence is defined as the total amount of radiant energy from all directions on a small sphere of cross-sectional area. Fluence is expressed in the unit of J/m^2 (Koutchma et al., 2009).

There are few devices that can be used to measure the UV dosage. Radiometer (thermal or photonic) is widely used to measure UV irradiance. The UV dose usually used for processing of pineapple juice was in the range of 5.6-53.42 mJ/cm^2 (Sew et al., 2014; Shamsudin et al., 2013; Chia et al., 2012; Goh et al., 2012). The wide range of UV dose used depends on nature of the pineapple juice such as fruit varieties or processes used (clarified or cloudy). Such differences will affect absorption coefficient and thus result in varying dose. Chemical actinometry can also be used to determine UV dose (Guerrero-Beltrán & Barbosa-Cánovas, 2004). Actinometers measure the amount of UV based on the concentrations of products that produced during photochemical reactions (Shama, 1999). Some of the common actinometers used during UV processing are potassium ferrioxalate ($K_3Fe(C_2O_4)_3$) and potassium iodide (KI). These actinometers have been utilized to measure the UV dose of dilute liquid but no report documented for the application of these actinometers on liquid with low pH and high concentrations of soluble solids, insoluble solids, and UV-absorbing substances such as juice. Another actinometer, 4,4',4"-tris-di-B-hydroxyethyaminotriphenyl-acetonitrile (HHEVC) has been studied for dose measurement during UV processing of apple juice. With the use of HHEVC dye, UV dose is defined as the received photons from a UV lamp that interacts with actinometer, which leads to chemical alteration at a particular wavelength (Koutchma et al., 2009).

Biodosimetry is the most consistent technique for measuring irradiance, which involves the measurement of log reduction of microorganisms after UV-C exposure under a particular condition (Sastry et al., 2000). In the biodosimetry application, the liquid product inoculated with microorganisms is flown through the UV reactor. The microbial inactivation is calculated by comparing the amount of microorganisms in influent and effluent of the reactor. *Escherichia coli* K2 can be used as the bacteria for inoculation in juice as it is a surrogate of *E. coli* O157:H7. The dose quantity is defined as the reduction equivalent dose. In order to obtain accurate dose measurement, all the microorganisms should deliver the same dose (ideal reactor) (Koutchma et al., 2009).

REGULATORY ASPECT

In 2000, FDA approved the use of UV irradiation as the alternative technique to thermal pasteurization of juices. The approval is listed in the FDA Code of Federal Regulations in Title 21 under Part 179.39 (21 CFR Part 179.39). FDA has investigated the safety of the application of UV irradiation for microbial inactivation of juices. The safety was evaluated according to the knowledge of the UV effects on the major chemical components present in food. From the findings, FDA concluded that there are no toxicological issues for any photochemical changes that may happen due to UV irradiation. The UV treatment suggested by FDA was effective to inactivate human pathogens in juices. There are many factors affecting the efficiency of UV irradiation, for examples type of juice, initial microbial load, and design of irradiation system (e.g., flow rate, number of lamps, and irradiation time). In this context, FDA did not define a specific dose limit by regulation, but the dose should be used according to good manufacturing practice. FDA expects the maximum dose will be limited in term of economic due to the cost of UV irradiation and organoleptic properties changes (e.g., changes in taste or color) that will affect consumer acceptance. However, FDA stated that 5 log reduction in pertinent microorganisms and pathogens must be achieved either through UV light treatment alone or in combination with several processes (21CFR Part 101.17). Besides, FDA specifies that the turbulent flow through tubes should be with a minimum Reynolds number of 2,200 for UV treatment of juice products (21 CFR Part 179.39).

In Canada, evaluation has been conducted by the Department of Novel Foods, of Health Canada on UV-treated apple juice/cider. The assessment focused on the microbial inactivation efficiency of *CiderSure* 3500 UV-light unit, comparison of composition and nutritional quality between UV light-treated apple juice/cider and untreated and pasteurized apple juice/cider, and the possibility of toxicological or chemical changes during UV treatment of apple juice/cider. The findings revealed that there was no human food safety concern for the UV light-treated apple juice/cider. The researchers proved that UV treatment is efficient to inactivate microorganisms in apple juice and cider products. Health Canada has approved the sale of apple juice and cider products treated with *CiderSure*3500 UV-light unit (Koutchma et al., 2009).

The European Union (EU) categorizes UV light as a type of irradiation. Regulations related to the application of irradiation in EU are still unclear. The EU still considers the type of food that allowed to be treated by ionizing radiation. Irradiation of food probably can be authorized provided that (i) there is no reasonable technological need, (ii) it does not cause health risk and it is conducted according to the conditions recommended, (iii) it brings benefits to consumer, and (iv) it is not applied as an alternative way for hygiene and health practice or good manufacturing or agricultural practice (Koutchma et al., 2009). At the moment, it is not known if other countries have established regulations on the application of UV irradiation for food processing.

BENEFITS OF UV

The benefits of using UV irradiation suggest the potential application of this technology in preserving fruit juices. UV irradiation has high inactivation efficiency against a wide range of microorganisms such as vegetative bacteria, bacterial spores, yeast, conidia (fungal spores), parasites, and viruses. Besides that, UV treatment also provides greater efficiency in maintaining quality and nutrients of juice compared to conventional thermal pasteurization treatment. Goh et al. (2012) found that UV irradiation was more effective than thermal pasteurization in preserving higher ascorbic acid content and antioxidant capacity in pineapple juice. Chia et al. (2012) also discovered that UV irradiation maintained better quality (total soluble solids, pH, titratable acidity, ascorbic acid, turbidity, total phenolic, and color) of pineapple juice compared to that treated with thermal pasteurization. Other than that, UV processing does not utilize chemicals or produce byproducts, making it to be an environmental-friendly technique. In addition, the small size of UV machine only requires a small floor area in the plant (Shah et al., 2016a; Mohd Adzahan & Benchamaporn, 2007). Furthermore, authorities such as FDA and Health Canada have approved the use of UV treatment for fruit juice processing.

Nutritional Content of Pineapple Juice

Pineapple fruit contains vitamin C, which can help the growth of body tissue, heal wounds, and assist injury discovery. Moreover, pineapple contains vitamin B1 (thiamin), vitamin B3 (niacin), potassium, sodium, and bromelain enzyme (MPIB, 2017). Carbohydrates in pineapple juice provide energy for human bodies. The potassium and sodium help to control the deep water balance in the human body. Meanwhile, pineapple fruit also contains bromelain enzyme, which aids in controlling swelling and infection as well as prevents blood clots (MPIB, 2017). Bromelain is sensitive to high temperature, therefore pasteurization of pineapple juice using nonthermal technology will be a better alternative. An attempt to pasteurize pineapple juice using UV technology without losing the activity of bromelain has been reported. Sew et al. (2014) reported that the application of UV maintained high activity of bromelain in pineapple juice.

EFFECTS OF UV RADIATION ON JUICE QUALITY

As mentioned in the previous section of this chapter, the efficiency of the UV technology relies on several factors such as UV-absorbing compounds (vitamins, antioxidants and others), color, turbidity, and presence of particulates. Another limiting factor is the initial microbial counts of fresh pineapple juice. The higher the initial counts, the more difficult it is for the UV light to penetrate. Therefore, the inactivation efficiency of UV will be reduced. Besides microbial inactivation, application of UV light is intended to preserve the quality of pineapple juice. Juice with higher acidity will have a longer shelf life especially when treated with UV.

EFFECT OF UV LIGHT ON MICROBIAL INACTIVATION IN JUICE

UV light inactivates microorganisms through disruption of their DNA or RNA structure. The primary mechanism of inactivation by UV is the creation of pyrimidine dimers, which are bonds formed between adjacent pairs of thymine or cytosine pyrimidines on the same DNA or RNA strand. The dimers prevent microorganisms from replicating, thereby rendering them inactive and unable to cause infection (Koutchma et al. 2009).

Among fruit juices, apple juice has been the most widely used medium to evaluate the microbial inactivation by UV light treatment. This is because apple juice is readily available, easily contaminated with microorganisms, and it can be produced as clear apple juice that can increase the depth of UV light penetration. Apart from that, other investigated juices include orange juice, tropical juice, guava juice, and pineapple juice. The target microorganisms for most of these investigations were *E. coli* O157:H7, *S. typhimurium* and *L.innocua* due to their high probability to cause food-borne illnesses. The ability of UV in inhibiting bacteria's replication is related to dimerization of their thymine bases in the DNA strands (Mukhopadhyay et al., 2012). In fruit juices, 90% of UV-C irradiation is absorbed in the first 1-mm from the surface (Keyser et al., 2008). Mansor et al. (2014) reported that UV irradiation at various dosages was effective in inactivating *S.typhimurium* in pineapple juice (Figure 15). The various effects on the juice quality are due to varying power levels, process times, UV-C light source-product distances, and product thickness to achieve varied inactivation levels of pathogens. Table 4 lists the effects of UV irradiation on microflora in pineapple juice.

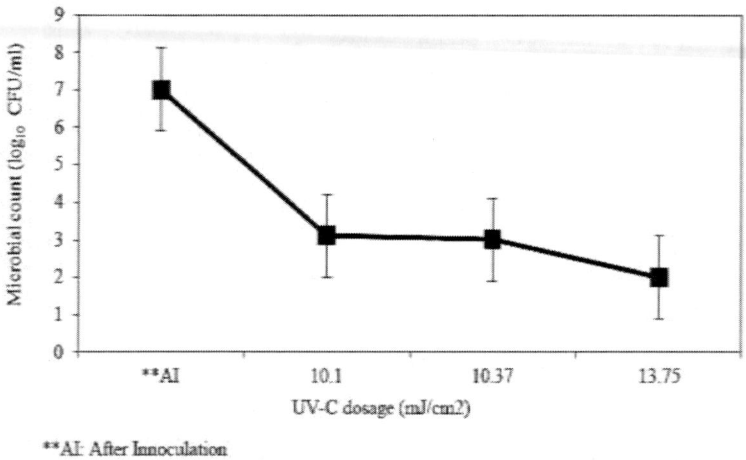

Figure 15. The average log_{10} count of *S. typhimurium* TISTR 292 in pineapple fruit juice after exposure to various UV-C dosages. Adapted from Mansor et al. (2014).

Table 4. Effects of UV irradiation on microflora in pineapple juice

Microflora	Dosage	Log reduction	References
L. innocua	5.61 mJ/cm^2+ mild heat (55°C for 10 min)	>5	Sew (2016)
Aerobic plate count Yeast and mold	10.76 mJ/cm 10. 76 mJ/cm	1.9 1.4	Unluturk et al. (2015)
S. typhimurium	0.000154 L/s	3.0	Mansor et al. (2014)
Aerobic plate count Yeast and mold	1607.0 J/L 1607.0 J/L	< 1.0 < 1.0	Keyser et al. (2008)

EFFECT OF UV IRRADIATION ON QUALITY AND SHELF LIFE OF PINEAPPLE JUICES

Various studies have highlighted the benefits of using UV irradiation with regard to quality and shelf life stability of juices. In the past twenty years, the usage of UV light was well documented with samples ranging from citrus juice to tropical juices. Table 5 highlights the effects of UV irradiation on quality of pineapple juice. Certain properties were negatively affected, certain properties were not affected, and certain properties were positively affected by UV treatment. Further details of the properties of UVtreated pineapple juice are described in the following sections.

Table 5. Effects of UV irradiation on quality of pineapple juice

Variables	Effect		References
	Negative	No effect/Positive	
UV dose = 10.76 mJ/cm^2	Increased turbidity, decreased vitamin C, and phenolic content	Retained pH, total soluble solids content, and titrable acidity	Shamsudin et al. (2014)
UV dose = 0.00154 L/s for 20-35s	NIL	Retained pH, total soluble solids content, ascorbic acid content, L* value.	Mansor et al. (2014)
UV dose = 5.61 mJ/cm^2 combined with mild heat (55°C for 10 min)	NIL	Retained color, bromelain activity, antioxidant capacity and TPC than thermal. Shelf life increased to 18 days.	Sew (2016)
UV dose = 5.61 mJ/cm^2 combined with mild heat (55°C for 10 min)	NIL	Decreased PME activity, retained bromelain content and TPC.	Sew et al. (2014)

Table 5. (Continued)

Variables	Effect		References
	Negative	No effect/Positive	
UV dose = 10.76 mJ/cm²	NIL	Retained viscosity and rheological properties	Shamsudin et al. (2013)
UV dose = 7.5 mJ/cm²	Decreased total soluble solids and titrable acidity, increased in turbidity.	Maintained total phenolic content until the fifth week of storage.	Chia et al. (2012)
UV dose = 53.42 mJ/cm²	Decreased ascorbic acid, carotenoids, phenolic acids, and antioxidant capacity, decreased total phenolic contents and flavonoids	Increased carotenoids, until 14 days of storage	Goh et al. (2012)

pH and Titratable Acidity

Acidity plays an important role in the sensory quality of fruit juice due to its flavor and astringency (Aguilar-Rosas et al., 2007). Studies by Mansor (2015) and Chia (2011) reported that the citric acid in UV-irradiated pineapple juice was similar to that of fresh juice. Figure 16 shows the changes in pH and titratable acidity of untreated, UV-irradiated, and thermally pasteurized pineapple juice. The exposure to UV light did not affect the sensory attributes of the juice and its quality remained similar to the untreated juice. On contrary, higher citric acid content was observed in thermally pasteurized pineapple juice. The increase in the acidity of thermally pasteurized juice was concomitant with the decrease in pH value. Leoni (2012) stated that organic acids are formed while sugars are decomposed during the heating process. These factors may explain the increased acidity in the juice during thermal pasteurization.

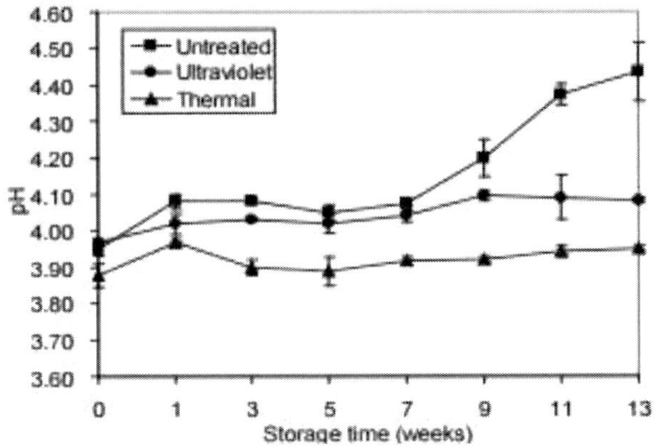

(A)

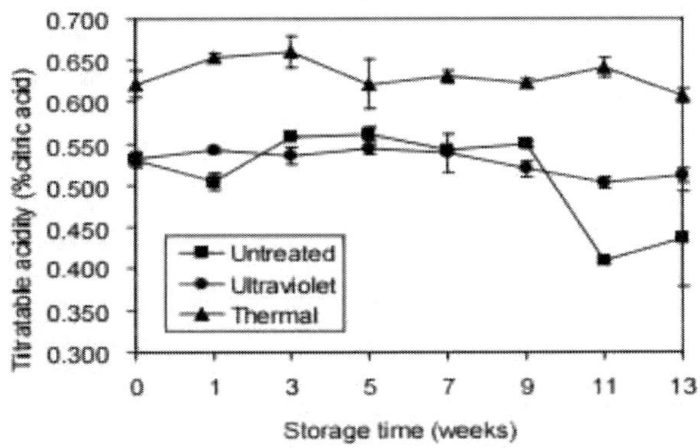

(B)

Figure 16. Changes in the pH (a) and titratable acidity (b) of untreated, ultraviolet-irradiated and thermally pasteurized pineapple juice. Adapted from Chia et al. (2012).

Vitamins

Spike et al. (1981) stated that vitamin A, B2, B12, D, E, K, carotenes, folic acid, tryptophan, and unsaturated fatty acids are 'light-sensitive' and can degrade when expose to UV-C light. Apart from that, vitamin C was the strongest absorber of UV light. A study by Koutchma et al. (2009) revealed that the reduction of vitamin C content was the cause of lessened pigmentation in the juice. The vitamin C content of fresh pineapple juice was reported to range from 9.2 to 93.8 mg/100 mL (Hounhouigan et al., 2014).

According to Shamsudin et al. (2014), loss of vitamin C content in the repetitive UV-irradiated and UV–dimethyl dicarbonate–UV treated pineapple juice was considered minimal compared to the pasteurized juice. Hence, UV irradiation provides a better alternative for juice treatment in order to preserve its attributes (color, flavor, and odor). Significant reduction in vitamin C was also observed in other fruit- juices such as apple juice (Falguera et al., 2011) and starfruit juice (Bhat et al., 2011). According to Davey et al. (2000), the decrease in the vitamin C content can be attributed to the oxidation process together with the activities of ascorbate oxidase and peroxidase enzymes. However, Chia (2011) found that UV irradiation retained ascorbic acid content in juice more efficiently, which gives benefit to the overall juice quality than a thermal process. Similar observations were reported by Goh et al. (2012).

Color

Color is one of the most important qualities of juice products. Chia et al. (2012) observed a higher L* value (lightness) and a high intensity color (chroma) of UV- irradiated pineapple juice compared to thermally pasteurized juice. A similar finding was reported by Mansor et al. (2015) whereby there was no adverse effect of UV irradiation on the color of pineapple juice as compared to the thermally pasteurized juice. According to Rattanathanalerk et al. (2005), enzymatic browning is neglected in the thermal processing of pineapple juice because the enzymes that cause browning are sensitive to heat at a temperature greater than 50°C (Martinez & Whitaker, 1995). UV irradiated pineapple juice has lightness similar to fresh juice as there is no substantial heat applied during UV irradiation.

Enzymes

The quality of fruit juice and nutritional properties are dependent on the activity of the enzymes present in the fruit. The application of UV light affects the activity of certain enzymes in pineapple juice including polyphenol oxidase (PPO) and pectin methylesterase (PME). PPO is responsible for enzymatic browning in fruit juice. According to Terefe, Buckow & Versteeeg (2014), PPO catalyzed the oxidation of *o*-dihydroxy phenols to *o*-quinones, which further polymerized to form undesirable brown pigments. Apart from that, the cloudiness of juice can be affected by the presence of PME. This enzyme hydrolyses the ester bonds of pectin in juices, resulting in decreased cloud stability (Rouse & Atkins, 1952). PME is generally inactivated using conventional heat pasteurization. There were many studies investigated the effect of UV on PPO and PME activities in fresh apple and orange juices. Noci et al. (2008) reported that the PPO activities in apple juice were not affected by UV treatment. Besides, Tran & Farid (2004) reported that there was no reduction of PME activity in orange juice by UV light at 78.3 mJ/cm^2. Similar finding was observed in the pineapple juice where Sew et al. (2014) showed that PME and proteolytic activities in freshly made pineapple juice were affected by mild heat treatment but not UV dose.

Turbidity

The turbidity of fruit juice is related to yeast and moulds (Donahue et al., 2005). Shamsudin et al. (2013) observed that the turbidity of pineapple juice decreased significantly after UV irradiation. The decrease in turbidity is due to the lower yeast and mould count after UV irradiation (Donahue et al. 2005). A similar finding was observed by Chia et al. (2012) (Figure 17) and Mansor et al. (2014). Meanwhile, Mohd Hanif et al. (2016) reported that there was no effect of UV treatment on the turbidity of cloudy tamarind juice.

In general, UV irradiation and thermal pasteurization are the effective technologies for destroying yeast and moulds, respectively. However, the turbidity of pineapple juice increased significantly after thermal pasteurization treatment. The degraded pectin formed during the heating process might combine with protein and hence increase the turbidity of the juice (Barros et al., 2004).

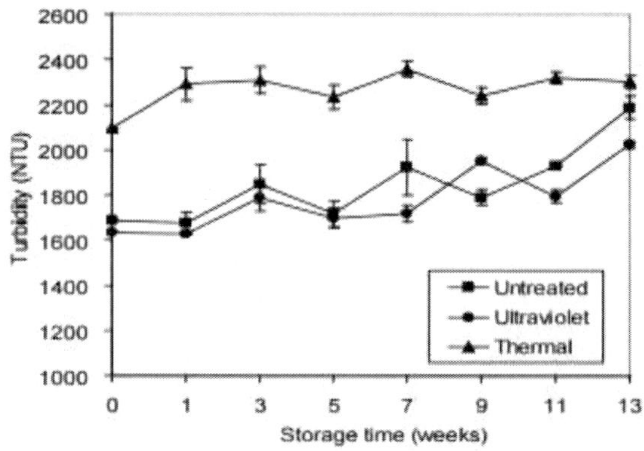

Figure 17. Changes in the turbidity of untreated, ultraviolet irradiated and thermally pasteurized pineapple juice. Adapted from Chia et al. (2012).

Antioxidant

Several studies reported that some antioxidants are destroyed by thermal treatment (Lessin et al., 1997; Lee and Coates, 2003; Cortés et al., 2006; Aguilar-Rosas et al., 2007; Gama and Sylos, 2007) and UV treatment (Pan et al., 2004; Koutchma, 2009). The losses of antioxidants may lead to colorless products. This was demonstrated by Chen et al. (1995) in carrot juice, which led to lower quality of products. Table 6 shows the antioxidant content and activity of thermal pasteurized and UV-irradiated pineapple juice.

Table 6. Antioxidant content and activity of pineapple juice. Adapted from Goh et al. (2012)

Parameter	Thermally pasteurized juice	UV-irradiated Juice
Ascorbic acid (mg/100g)	10.065 ± 0.798^{b}	12.673 ± 1.113^{c}
Carotenoids (ug/L)	5.084 ± 0.994^{b}	5.810 ± 0.866^{b}
Phenolic acids (mg GAE/100g)	41.810 ± 14.590^{a}	43.020 ± 4.380a
Flavonoids (mg CEQ/100g)	24.933 ± 6.890^{a}	20.021 ± 5.181a
Antioxidant Capacity (%)	19.807 ± 6.523^{b}	29.793 ± 4.991ab

Mean values having different superscript low case letter within the row are significantly different (p <0.05).

Ascorbic acid can also act as an antioxidant. Reduction of the ascorbic acid content of thermally treated juices was reported by several authors (Achinewhu & Hart, 1994; Iversen, 1999; Tiwari et al., 2009; Alothman et al., 2009). Davey et al. (2000) stated that the reduction of ascorbic acid content could be attributed to the oxidation process together with the activities of ascorbate oxidase and peroxidase enzymes, which can also affect the phenolic and antioxidant compounds in the UV-treated juice.

Phenolic compounds are secondary metabolites in plants that are important for flavor and color development in fruit juice. Phenolic compounds contribute to the characteristics of odor and flavor in apple products (Aguilar-Rosas et al., 2007) and might play the same functional role in pineapple juice. Phenolic compounds also provide antioxidant potential and health promoting properties (Kaur & Kapoor, 2001). Chia et al. (2012) reported that phenolic loss in pineapple juice treated with thermal pasteurization was higher compared to UV-irradiated juice, which the phenolic content was only slightly reduced. Goh et al. (2012) indicated similar findings in which phenolics were significantly lower after UV-C treatment, but were significantly higher than thermally treated samples. The changes of phenolic content in the juice may be related to peroxidase that degrades the phenolic compounds, leading to quality loss of juice (Odriozala-Serrano et al., 2008).

Sensory Quality

Sensory quality of pineapple juice has not been well explored. Sensory quality of fruit juices plays an important role in consumer satisfaction. Most studies reported that UV-C-treated juices were not significantly different than fresh juices but were significantly different from thermal treated juices. Hanisah (2009) reported that UV-irradiated pineapple juice was highly preferred compared to thermal pasteurized juice. The thermally pasteurized pineapple juice was significantly less preferred for odor, color, cloudiness, acidity, overall flavor, and overall likeness than the control and UV-C-treated samples (Choi et al., 2005; Hanisah et al., 2005). Current studies have focused on orange and apple juices as well as apple cider. Pala and Toklucu (2013) reported that UV processing of freshly squeezed orange juice at 48.1 kJ/L could assure its safety and improve its sensory attributes in comparison to heat treatment.

Shelf Life of UV Irradiated Pineapple Juice

Shelf life of UV-irradiated pineapple juice is highly dependent on the amount of polyphenol, ascorbic acid, and sugar left within the juice, coupled with the amount of pertinent microorganisms found in the fruit juice during standard 12-weeks storage. These are naturally decaying properties of juice kept at 4°C (standard chiller temperature). A study by Chia (2011) showed that UV irradiation preserved the physicochemical and rheological properties of pineapple juice thus prolonged the shelf life at least 6 weeks longer than untreated pineapple juice. Moreover, Mansor (2015) proved that the shelf life of pineapple juice was extended up to 5 weeks. When UV treatment was combined with mild heat (55°C for 10 min), the shelf life of pineapple juice was extended up to 18 days (Sew, 2016).

Hurdle Technology of UV Irradiated Pineapple Juice

Food and Drug Administration (FDA) stated that 5 log reduction in microorganisms and pathogens must be achieved for UV light treatment of fruit and vegetable juice. UV irradiation on juice is not sufficient to achieve the maximal reduction of microorganisms. Due to the limitation of UV irradiation, hurdle technology with other processing techniques could achieve the required limit in microbial reduction and retention of juice quality. The combination could be between UV technology with heat or it could be between UV irradiation with other nonthermal methods or application of additives or preservatives. Pineapple juice has been tested using hurdle technology by combining UV treatment with dimethyl dicarbonate and mild heat, respectively (Shamsudin et al., 2014; Sew et al., 2014). Dimethyldicarbonate (DMDC) is one of the effective anti-microbial agents that can control a wide range of microorganisms, which include *E. coli* O157:H7 and yeast. According to Shamsudin et al. (2014), UV-C treatment (10.76 mJ/cm^2) and dimethyl dicarbonate (DMDC250 ppm) significantly reduced the microbial count in pineapple juice. Furthermore, turbidity, vitamin C, and phenolic content were found to have significant changes.

Pineapple juice has beneficial health compounds such as bromelain and phenols (Hale, Greer, Trinh, & James, 2005) and also possess antioxidant activity (Hossain & Rahman, 2011). Conventional heat pasteurization compromises the heat sensitive phenolic compounds (Goh, Noranizan, Leong, Sew, & Sobhi, 2012) and bromelain (Bhattacharya & Bhattacharya, 2009) as well as the quality of the juice (Rattanathanalerk et al. 2005; Shamsudin et al., 2014; Shamsudin et al., 2013). Bhattacharya & Bhattacharya (2009) reported that devoid of bromelain led to the absence of proteolytic activity in pineapple products. Recently, Sew et al. (2014) reported that the combined treatment of mild heat at 55°C for 10 min followed by UV treatment at 5.61 $mJ \cdot cm^{-2}$ was recommended to produce pineapple juice with relatively high contents of bromelain and total phenolic but low PME activity.

POTENTIAL COST IMPLICATION OF UV IRRADIATED PINEAPPLE JUICE

Pineapple juice is one of the products produced in canned pineapple industry. Most of the equipment necessary for pineapple juice production are similar to that for the production of canned pineapples. Thus, low equipment costs are needed for producing pineapple juice (Lau et al., 2011). Table 7 indicates that UV pasteurization could produce a cheaper pineapple juice compared to thermal pasteurization. Higgins (2003) and Kozempel et al. (1998) reported similar cost for UV-pasteurized (approximately RM 1.60 per 100 liters) and thermally pasteurized apple cider (approximately RM 4.00 per 100 liters), respectively.

Table 7. Comparison of cost-volume-profit analysis for canned pineapple juices. Adapted from Lau et al. (2011)

Parameter	UV-treated pineapple juice	Thermally treated pineapple juice
Variable cost per can (320 ml)	RM 0.895	RM0.900
Contribution margin per unit	RM 0.605	RM 0.600
Break-even point	567.273 cans per year	712,000 cans per year
Margin of safety	67%	62%

Higher selling price could be set for the products that have premium quality such as products with better color and flavor profile, higher nutrient retention as well as fresh-like characteristics. Hanisah (2009) reported that better quality of pineapple juice (good taste, color profile, and ascorbic acid content similar to that of freshly pressed juice) was obtained when treated with UV pasteurization.

Figure 18 illustrates the positioning of UV-treated pineapple juice in terms of perceived quality and price. Positioning maps were used to identify consumer perceptions of their brands among other competing products on important buying dimensions (Tan et al., 2015). Through the positioning maps, UV-irradiated pineapple juice was marked as highly preferred compared to the other commercial brands of thermally treated juice. Shugan (1984) stated that the price of UV-irradiated juice was relatively high due to its high perceived quality. As consumers willing to pay more for UV-treated juice, it could be beneficial for small scale manufacturers. In addition, the implementation of UV technology can be more profitable when applied in a small-to-medium scale pineapple juice setup and thus a plus point for small scale manufacturers (Shah et al., 2016a).

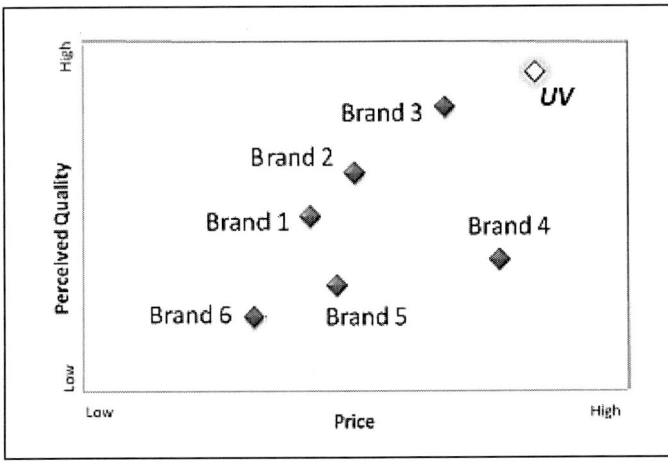

(A)

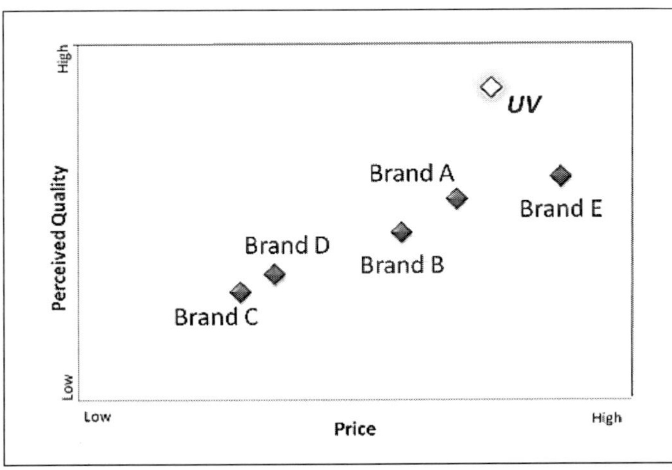

(B)

Figure 18. Positioning map for pineapple juice (a) and single strength fruit juice products (b) in Malaysia. Adapted from Tan et al. (2015).

FUTURE OF UV IRRADIATED PINEAPPLE JUICE

The basic designs of UV equipment have been explored and patented. Based on studies of market and consumer survey as well as cost implication as mentioned above, the potential to market UV-irradiated pineapple juice as a premium juice is realistic and attractive. A lot of research have been carried out for the effects of UV irradiation on quality (physicochemical properties, functional properties, rheology, and sensory) and shelf life of pineapple juice. Research was also carried out for UV treatment with various doses and for different pineapple species and juice blends (Sew et al., 2014; Mansor et al., 2014; Chia et al., 2012). The effectiveness of hurdle technology combining UV irradiation with dimethyl dicarbonate or mild heat on maintaining the quality of pineapple juice was also explored (Shamsudin et al., 2014; Sew et al., 2014). Table 8 shows the summary of areas related to UV irradiation of pineapple juice, which have been explored and currently being/yet to be explored.

Despite the areas explored, there are still knowledge gaps in term of process, analysis, and commercialization to be filled for UV processing of pineapple juice. Although regulations of UV irradiation have been established in developed countries, the legislations regarding the use of UV irradiation for the preservation of juice should be developed. The application of hurdle technology using UV irradiation in combination with other processes such as ultrasound, membrane technology, addition of additives, and gas treatment is potential to further improve the quality of pineapple juice. Other than hurdle technology, the application of high intensity pulsed UV on shelf life extension of pineapple juice is an interesting topic to be explored. In terms of analysis, the study on flavor, aroma, and organic acid content of pineapple juice is still lacking. Aroma and organic acid content of pineapple juice can be examined using gas chromatography-mass spectrometry and high performance liquid chromatography, respectively. Several researchers have highlighted the effect of UV irradiation on microorganisms such as *E. coli* O157:H7 and *S. typhimurium* but there is still limited knowledge of its effect on viruses, parasites, and spores. Other than that, furan is a carcinogen and present in thermally-treated foods. Since UV irradiation is an alternative method to thermal pasteurization, the presence of furan in UV-irradiated juice should be investigated in order to ensure the safety of the juice. Furan development was observed by Shah et al. (2016b) for pummelo fruit juice.

Table 8. Summary of areas related to ultraviolet irradiation of pineapple juice which has been explored and yet to be explored

ULTRAVIOLET IRRADIATION OF PINEAPPLE JUICE					
AREAS EXPLORED			AREAS NOT YET OR CURRENTLY BEING EXPLORED		
Process	Analysis	Commercialization	Process	Analysis	Commercialization
• Hurdle process: (mild heat and dimethyl dicarbonate) • Dose measurement • Regulations (mostly in developed countries) • Various pineapple species and juice blends	• Shelf life • Physico-chemical properties • Functional properties • Rheology • Sensory • Market and consumer survey • Cost implication	• Basic designs • Patent • Automated, self-adjustable flow rate	• Hurdle process: ultrasound, membrane technology, additives, gasses • Regulations (in other countries) • High Intensity Pulsed UV	• Flavor and aroma • Furan products • Organic acid • Virus, parasite, spore	• Scale-up • Market trial • High performance and high volume unit • Thermal assisted UV unit

However, furan content in UV-irradiated pineapple juice is not well-reported and it should be investigated. Besides process and analysis, commercialization aspect of UV-irradiated pineapple juice is worth for further study in order to commercialize it as premium juice. This can be done by conducting scale-up and market trial as well as the development of high performance and high volume unit and thermal-assisted UV unit. Knowledge on UV processing of pineapple juice will be more complete with the study in all of these areas that are yet to be explored.

In brief, UV light technology as an alternative preservation technique to thermal pasteurization is highly promising in extending the shelf life and maintaining the quality of pineapple juice. In this context, the commercialization of UV-irradiated pineapple juice by the food industry as a premium quality juice is possible and can be profitable.

REFERENCES

Achinewhu, S. C. and Hart, A. D. (1994). Effect of processing and storage on the ascorbic acid (vitamin C) content of some pineapple varieties grown in the Rivers State of Nigeria. *Plant Foods for Human Nutrition* 46: 335-337.

Aguilar-Rosas, S. F., Ballinas-Casarrubias, M. L., Nevarez-Moorillon, G. V., Martin-Belloso, O. and Ortega-Rivas, E. (2007). Thermal and pulsed electric fields pasteurization of apple juice: Effects on physicochemical properties and flavour compounds. *Journal of Food Engineering*, 83: 41-46.

Al-Azzawi, A. (2006). *Light and Optics: Principles and Practices*. Florida: CRC Press.

Alothman, M., Bhat, R. and Karim, A. A. (2009). Antioxidant capacity and phenolic content of selected tropical fruits from Malaysia, extracted with different solvents. *Food Chemistry*, 115: 785–788.

Bandla, S., Choudhary, R., Watson, D. G. and Haddock, J. (2012). UV-C treatment of soymilk in coiled tube UV reactors for inactivation of

Escherichia coli W1485 and *Bacillus cereus* endospores. *Food Science and Technology,* 46: 71–76.

Barros, S. T. D., Andrade, C. M. G., Mendesa, E. S. and Peres, L. (2003). Study of fouling mechanism in pineapple juice clarification by ultrafiltration. *Journal of Membrane Science*, 215: 213–224.

Bhattacharya, R. and Bhattacharya, D. (2009). Preservation of natural stability of fruit "bromelain" from *Ananas comosus* (pineapple). *Journal of Food Biochemistry*, 33 (1):1–19.

Bates, R. P., Moris, J. R. and Crandall, P. G. (2001). *Principles and practices of small and medium- scale fruit juice processing*. FAO Agriculture Services Bulletin, 146: 135-149.

Bhat, R. Ameran, S. B., Voon, H. C., Karim, A. A. and Tze, L. M. (2011). Quality attributes of starfruit (*Averrhoa carambola* L.) juice treated with ultraviolet radiation. *Food Chemistry,*127 (2): 641–644.

Bohren, C. F. and Huffman, D. R. (2004). *Absorption and Scattering of Light by Small Particles.* Weinheim: WILEY-VCH.

Butcher, G. (2016). *Tour of the Electromagnetic Spectrum*. Washington: NASA.

Chia, S. L. (2011). "Effect of Ultraviolet Irradiation on Physicochemical, Microbial and Rheological Properties of Pineapple (*Ananas Comosus L.* Var. Yankee) Juice." Master diss., University Putra Malaysia.

Chia, S. L., Rosnah, S., Noranizan, M. A. and Wan Ramli, W. D. (2012). The effect of storage on the quality attributes of ultraviolet-irradiated and thermally pasteurised pineapple juices. *International Food Research Journal*, 19 (3): 1001-1010.

Choi, L. H. and Nielsen, S. S. (2005). The effects of thermal and nonthermal processing methods on apple cider quality and consumer acceptability. *Journal Food Quality*, 28: 13–29.

Code of Federal Regulations (CFR). (2001). Important of fruits and vegetables, 7 CFR Part 300 and 319. Office of Federal Register: *Rules and Regulations*, 66 (167): 45151-45161. Washington :Government Printing Office.

Cortés, C., Esteve, M. J., Rodrigo, R., Torregrosa, F. and Frígola, A. (2006). Changes of colour and carotenoids content during intensity pulsed electric field treatment in orange juices. *Food and Chemistry Toxicology*, 44: 1932-1939.

Davey, M. W., Van Montagu, M., Inze, D., Sanmartin, M., Kanellis, A., Smimoff, N., Benzie, L. J. J., Strain, J. J., Favell, D. and Fletcher, J. (2000). Plant L-ascorbic: Chemistry, function, metabolism, bioavailable and effects of processing. *Journal Science and Food Agriculture,* 80: 825–860.

Donahue, D. W., Canitez, N. and Bushway, A. A. (2004) UV-C inactivation of *E. coli* O157:H7 in apple cider: Quality, sensory & shelf-life analysis. *Journal Food Processing and Preservation.,* 28: 368–387.

Donahue, D. W, Canitez, N. and Bushway, A. A. (2005). UV inactivation of *E.coli* O157:H7 in 360 apple cider: quality, sensory and shelf life analysis. *Journal of Food Processing and Preservation*, 28(5): 368-387.

Elez-Martínez, P. and Martin-Belloso, O. (2005). "Food Safety Aspects of Pulsed Electric Fields." In *Emerging technologies for food processing*, edited by D.W. Sun, 183-218. Boston: Academic Press.

EPA (U.S. Environmental Protection Agency). (2006). "*Ultraviolet Disinfection Guidance Manual for the Final Long Term 2 Enhanced Surface Water Treatment Rule.*" Accessed February 10. https://nepis.epa.gov/Exe/ZyPDF.cgi/600006T3.PDF?Dockey=600006T3.PDF.

Falguera, V., Pag´an, J. and Ibarz, A. (2011). Effect of UV irradiation on enzymatic activities and physicochemical properties of apple juices from different varieties. *LWT - Food Science and Technology*, 44 (1): 115–119.

FDA. (2015). *"Kinetics of Microbial Inactivation for Alternative Food Processing Technologies – Ultraviolet Light."* Accessed February 24. https://www.fda.gov/Food/FoodScienceResearch/SafePracticesforFoodProcesses/ucm103137.htm.

FDA. (2016). "*CFR – Code of Federal Regulations Title 21. Part 101.17 – Food labelling warning, notice, and safe handling statements*."

Accessed March 1. https://www.accessdata.fda.gov/scripts/cdrh/cfdocs/cfcfr/CFRSearch.cfm?fr=101.17.

FDA. (2016). "*CFR – Code of Federal Regulations Title 21. Part 179 – Irradiation in the production, processing and handling of food.*" Accessed February 25. https://www.accessdata.fda.gov/scripts/cdrh/cfdocs/cfCFR/CFRSearch.cfm?fr=179.39.

Gama, J. J. T. and Sylos, C. M. (2007). Effect of thermal pasteurization and concentration on carotenoid composition of Brazilian Valencia orange juice. *Food Chemistry*, 100: 1686-1690.

Goh, S. G., Noranizan, M., Leong, C. M., Sew, C. C. and Sobhi, B. (2012). Effect of thermal and ultraviolet treatments on the stability of antioxidant compounds in single strength pineapple juice throughout refrigerated storage. *International Food Research Journal*, 19: 1131–1136.

Gómez-López, V. M., Ragaert, P., Debevere, J. and Devlieghere, F. (2007). Pulsed light for food decontamination: A review. *Trends in Food Science and Technology,* 18 (9): 464-473.

Guerrero-Beltrán, J. A. and Barbosa-Cánovas, G. V. (2004). Review: advantages and limitations on processing foods by UV light. *Food Science Technology International,* 10 (3): 0137-0147.

Hanisah, H. (2009). "Effect of Ultraviolet Irradiation on the Quality of Pineapple Juice." Bachelor diss., Universiti Putra Malaysia.

Hale, L. P., Greer, P. K., Trinh, C. T. and James, C. L. (2005). Proteinase activity and stability of natural bromelain preparations. *International Immunopharmacology*, 5 (4): 783–793.

Higgins, K. T. (2003). Fresh Today, Safe Next week. *Food Engineering Magazine*. Retrieved 21 February 2017 from http://www.foodengineeringmag.com/articles/84458-fresh-today-safe-next-week.

Hossain, M. A. and Rahman, S. M. M. (2011).Total phenolics, flavonoids and antioxidant activity of tropical fruit pineapple. *Food Research International*, 44: 672-676.

International Tropical Fruits Network. (2011). "Pineapple: processing technical information." Accessed December 23. http://www.ifnet.org/gfruit/Templates%20English/pineapple.process.info.htm.

Iversen, C. K. (1999). Blackcurrant nectar: Effect of processing and storage on anthocyanin and ascorbic acid content. *Journal of Food Science*, 64: 37-41.

Jawad, M. M., Qader, S. T. A., Zaidan, A. A., Zaidan, B. B., Naji, A. W. and Qader, I. T. A. (2011). An overview of laser principle, laser-tissue interaction mechanisms and laser safety precaution for medical laser users. *International Journal of Pharmacology*, 7 (2): 149-160.

Kaur, C. and Kapoor, H. C. (2001). Antioxidants in fruits and vegetables-the millenium's health. *International Journal of Food Service and Technology*, 36: 703-725.

Keyser, M., Muller, I. A., Cilliers, F. P., Nel, W. and Gouws, P. A. (2008). Ultraviolet radiation as a non-thermal treatment for the inactivation of microorganisms in fruit juice. *Innovative Food Science Emerging Technology*, 9: 348–354.

Kipphan, H. (2001). *Handbook of print media: Technologies and production methods.* Berlin: Springer-Verlag.

Koutchma, T., Popovic, V., Ros-Polski, V. and Popielarz, A. (2016). Effects of ultraviolet light and high-pressure. *Comprehensive reviews in Food Science and Food Safety,* 15: 844-867.

Koutchma, T. (2009) Advances in ultraviolet light technology for non-thermal processing of liquid foods. *Food Bioprocess Technology*, 2: 138–155.

Koutchma, T. (2008). "*UV light for processing foods*." Accessed February 25. https://pdfs.semanticscholar.org/1418/b4df34a14e4eda375e96c905d35cfcd43628.pdf.

Koutchma, T. N., Forney, L. J. and Moraru, C. I. (2009). *Ultraviolet light in food technology*. Florida: CRC Press.

Kozempel, M., McAloon, A. and Yew, W. (1998). The cost of pasteurizing apple cider. *Journal Food Technology,* 52: 50–52.

Loeillet, D. (2006). "*The international pineapple trade: The great year.*" Accessed December 13. http://passionfruit.cirad.fr/index.php/download/%28id%29/2685/%28langue%29eng/%28type%29/article.

Lau, P. L., Mohd Adzahan, N., Hashim, N., Shamsudin, R., Sew, C. C. and Sobhi, B. (2011) Pineapple juice production using ultraviolet pasteurisation: Potential cost implications. *Journal Agribusiness Marketing*, 4: 38–50.

Lee, H. S. and Coates, G. A. (2003). Effect of thermal pasteurization on Valencia orange juice color and pigments. *Food Science Technology*, 36: 153–156.

Lessin, W. J., Catignani, G. L. and Schwartz, S. J. (1997). Quantification of cis-trans isomers of provitaminA carotenoids in fresh and processed fruits and vegetables. *Journal of Agricultural and Food Chemistry*, 45: 3728-3732.

Lu, G., Li, C., Liu, P., Cui, H., Xia, Y. and Wang, J. (2010). Inactivation of microorganisms in apple juice using an ultraviolet silica-fiber optical device. *Journal of Photochemistry and Photobiology*, 100: 167–172.

Mansor, A., Shamsudin, R., Mohd Adzahan, N. and Hamidon, M. N. (2014). Efficacy of ultraviolet radiation as a non-thermal treatment for the inactivation of *Salmonella typhimurium* TISTR 292 in pineapple fruit juice. *Agriculture and Agricultural Science Procedia*, 2: 173–180.

Mansor, A., Shamsudin, R., Mohd Adzahan, N. and Hamidon, M. N. (2015). Performance of UV pasteurization with quartz glass sleeve on physicochemical properties and microbial activity of pineapple juice. *Journal of Food Processing Engineering*, 40 (1): e12263.

McDonald, K. F., Curry, R. D., Clevenger, T. E., Unklesbay, K., Eisenstark, A., Golden, J. and Morgan, R. D. (2000). Comparison of pulsed and continuous ultraviolet light sources for the decontamination of surfaces. *IEEE Transactions on Plasma Science,* 28 (5): 1581-1587.

Malaysian Pineapple Industry Board (MPIB). (2017). *"Nutrition of Pineapple."* Accessed February 26. http://www.mpib.gov.my/en /khasiat-nanas.

Mohd Adzahan, N. (2006). "Effects of Ultraviolet Treatmenton Water Soluble Vitamin Retention in Aqueous Model Solutions and Apple Juice." PhD diss., Cornell University.

Mohd Adzahan, N. and Benchamaporn, P. (2007). Potential of non-thermal processing for food preservation in Southeast Asian countries. *ASEAN Food Journal*, 14(3): 141-152.

Mohd-Hanif, H. A., Shamsudin, R. and Mohd Adzahan, N. (2016). Effects of UVC irradiation and thermal treatment on the physico-chemical properties and microbial reduction of clear and turbid tamarind juice. *International Food Research Journal*, 23: 107-112.

Moraru, C. I. and Uesugi A. (2009). "Pulsed light treatment: Principles and applications." In *Ultraviolet light in food technology: Principles and applications*, edited by T. Koutchma, L. Forney, and C. I. Moraru, 235-265. Florida: CRC Press.

Mukhopadhyay, S. and Ramaswamy, R. (2012). Application of emerging technologies to control Salmonella in foods: A review. *Food Research International,* 45: 666–677.

Müller, A., Stahl, M. R., Graef, V., Franz, C. M. A. P. and Huch, M. (2011). UV-C treatment of juices to inactivate icroorganisms using dean vortex technology. *Journal of Food Engineering,* 107: 268–275.

Müller, A., Noack, L., Greiner, R., Stahl, M. R. and Posten, C. (2014). Effect of UV-C and UV-B treatment on polyphenoloxidase activity and shelf life of apple and grape juices. *Innovative Food Science Emerging Technology*, 26: 498–504.

Murakami, E. G., Jackson, L., Madsen, K. and Schickedanz, B. (2006). Factors affecting the ultraviolet inactivation of Escherichia coli K12 in apple juice and a model system. *Journal of Food Process Engineering*, 29, 53–71.

Nemeth, J. and Bucsky, G. (1997). Secondary flow generated in curved flow. *Hungarian Journal of Industrial Chemistry*, 25: 91-98.

Neil, K. P. (2011). *"Medical detectives: Foodborne disease surveillance and outbreak detection."* Accessed January 20. http://www.asm.org /branch/brcano/Neil.pdt.

Noci, F., Riener, J., Walkling, M., Cronin, D. A., Morgan, D. J. and Lying, J. G. (2008). Ultraviolet irradiation and pulsed electric fields (PEF) in a hurdle strategy for the preservation of fresh apple juice. *Journal of Food Engineering*, 85: 141–146.

Noranizan, M. A., Lau, P. L., Narimah, H., Rosnah, S., Sew, C. C. and Babak, S. (2011). Pineapple juice production using ultraviolet pasteurisation: Potential cost implications. *Journal of Agribusiness Marketing*, 4: 38-50.

Noranizan, M., Sharizah, S., Sew, C. C. and Karim, R. (2011). "Comparison between pulsed light and ultraviolet treatment on microflora survival and quality attributes of pineapple juice." Paper presented at the Universiti Malaysia Terengganu 10th International Annual Symposium (UMTAS), Kuala Terengganu, Malaysia, July 11-13.

Pala, C. U. and Toklucu, A. K. (2013). Microbial, physicochemical and sensory properties of UV-C processed orange juice and its microbial stability during refrigerated storage. *LWT - Food Science Technology*, 50 (2): 426–31.

Pan, J., Vicente, A. R., Martinez, G. A., Chaves, A. R. and Civello, P. M. (2004). Combined use of UV-C irradiation and heat treatment to improve postharvest life of strawberry fruit. *Journal of the Science of Food and Agriculture*, 84: 1831-1838.

Pataro, G., Muñoz, A., Palgan, I., Noci, F., Ferrari, G. and Lyng, J. G. (2011). Bacterial inactivation in fruit juices using continuous flow pulsed light (PL) system. *Food Research International,* 44: 1642–1648.

Ramaswamy, H. and Marcotte, M. (2006). *Food Processing: Principle and Applications*. USA: CRC Press, Taylor & Francis Group.

Rattanathanalerk, M., Chiewchan, N. and Srichumpoung, W. (2005). Effect of thermal processing on the quality loss of pineapple juice. *Journal of Food Engineering*, 66: 259-265.

Rosnah, S., Noranizan, M. A. and Yap, P. Y. (2013). Ultraviolet technology- An alternative to juice pasteurization. *Jurutera: Food Security and quality in Malaysia*, 1: 26-30.

Rouse, A. H. and Atkins, C. D. (1952). Heat inactivation of pectin esterase in citrus juices. *Food Technology*, 6: 291–294.

Sabbe, S., Verbeke, W. and Damme, P. V. (2008). Familiarity and purchasing intention of Belgian consumers for fresh and processed tropical fruit products. *British Food Journal*, 110 (8): 805-818.

Sanchez-Vega, R., Mujica-Paz, H., Marquez-Melendez, R., Ngadi, M. O. and Ortega-Rivas, E. (2009). Enzyme inactivation on apple juice treated by ultra pasteurization and pulsed electric fields technology. *Journal of Food Processing and Preservation*, 33: 486-499.

Sastry, S. K., Datta, A. K. and Worobo, R. W. (2000). Ultraviolet light. *Journal of Food Science, Supplement*, 65 (12): 90-92.

Sew, C. C. (2016). "Nonthermal Treatments as Alternatives to Thermal Pasteurisation of Pineapple Juice." Master diss., Universiti Putra Malaysia.

Sew, C. C., Mohd Ghazali, H., Martín-Belloso, O. and Noranizan, M. A. (2014). Effects of combining ultraviolet and mild heat treatments on enzymatic activities and total phenolic contents in pineapple juice. *Innovative Food Science and Emerging Technology*, 26: 511–6.

Sew, C. C., Ghazali, H., Martín-Belloso, O. and Noranizan, M. (2012). "Effects of mid-thermal and ultraviolet treatments on Listeria innocua inactivation in pineapple juice." Paper presented at the International Nonthermal Food Processing Workshop, Melbourne, Australia, October 16–17..

Shah, N. N. A. K., Shamsudin, R., Rahman, R. A. and Adzahan, N. M. (2016a). Fruit juice production using ultraviolet pasteurization: A review. *Beverages*, 2: 22-41.

Shah, N. N. A. K., Shamsudin, R., Rahman, R. A. and Adzahan, N. M. (2016b). Furan development in dean vortex UVC treated pummel (*Citrus grandis* L. Osbeck) fruit juice. *International Food Research Journal*, 23 (Suppl): S113-S118.

Shama, G. (1999). "Ultraviolet Light." In *Encyclopedia of Food Microbiology-3*, edited by R. K. Robinson, C. Batt, and P. Patel, 2208-2214. London: Academic Press.

Sharma, R. R. and Demirci, A. (2003). Inactivation of *Escherichia coli* O157:H7 on inoculated alfalfa seeds with pulsed ultraviolet light and response surface modelling. *Journal of Food Science*, 68 (4): 1448-1453.

Shamsudin, R., Chia, S. L., Mohd Adzahan, N. and Wan Daud, W. R. (2013). Rheological properties of ultraviolet-irradiated and thermally pasteurized Yankee pineapple juice. *Journal of Food Engineering,* 116: 548–553.

Shamsudin, R., Noranizan, M. A., Yap, P. Y. and Mansor, A. (2014). Effect of repetitive ultraviolet irradiation on the physico-chemical properties and microbial stability of pineapple juice. *Innovative Food Science Emerging Technology,* 23: 114-120.

Spikes, J. (1981). *Photochemical and Photobiological Reviews Volume 6.* New York: USA.

Tan, H. Y., Ganesh, T. and Noranizan, M.A. (2014). "Market Potential Analysis and Possible Marketing Strategy for Ultraviolet-Irradiated Single Strength Pineapple Juice in the Klang Valley." In *Marketing: A Compendium*, edited by K. T. G. Cheng, H. J. Amer, and T. Ganesh, 82–105. Selangor: Universiti Putra Malaysia Press.

Terefe, N. S., Buckow, R. and Versteeg, C. (2014). Quality-related enzymes in fruit and vegetable products: effects of novel food processing technologies, part 1:high-pressure processing. *Critical Reviews in Food Science and Nutrition,* 54 (1):24–63.

Tiwari, B. K., Muthukumarappan, K., O'Donnell, C. P. and Cullen, P. J. (2009). Inactivation kinetics of pectin methylesterase and cloud retention in sonicated orange juice. *Innovative Food Science and Emerging Technologies*, 10: 166–171

Tran, M. T. T. and Farid, M. (2004) Ultraviolet treatment of orange juice. *Innovative Food Science and Emerging Technologies,* 5: 495–502.

Unluturk, S. and Atilgan, M. R. (2015).Microbial safety and shelf life of UV-C treated squeezed white grape juice. *Journal of Food Science*, 80: 1831–1841.

Zhang, C., Trierweiler, B., Li, W., Butz, P., Xu, Y., Rufer, C. E., Ma, Y. and Zhao, X. (2011). Comparison of thermal, ultraviolet-c, and high pressure treatments on quality parameters of watermelon juice. *Journal of Food Control*, 126: 254–260.

In: Tropical Fruits
Editors: C. Stewart Bogsan et al.
ISBN: 978-1-53612-885-7

Chapter 8

PINEAPPLE FRUIT: TECHNICAL ASPECTS OF CULTIVATION, POST-HARVEST AND NUTRITION

Henriqueta Talita Guimarães Barboza[1], Alexandra Mara Goulart Nunes Mamede[1], Antonio Gomes Soares[1,*], Otniel Freitas Silva[1], Valéria Bezerra Saldanha[2], José Ribamar Gusmão Araujo[3] and Augusto César Vieira Neves Junior[3]
[1]Embrapa Food Technology, Rio de Janeiro, Brazil
[2]Embrapa Amapá, Amapá, Brazil
[3]Universidade Estadual do Maranhão, Maranhão, Brazil

ABSTRACT

Fresh fruits are excellent sources of energy, vitamins, minerals and fibers. Pineapple (*Ananas comosus*) is the 3rd most important fruit traded

[*] Corresponding Author Email: antonio.gomes@embrapa.br.

globally and one of the most popular and delicious tropical fruits due its good sensorial characteristics such as mouth feeling, flavor, acidity/sweetness ratio, color and nutrition. This fruit is also considered as a rich Data from of vitamins B and C, besides several minerals and other health benefits. Pineapple is also a Data from of bromelain used as a food supplement because of its many phytomedicinal properties. Pineapple contributes to over 20% of the tropical fruit produced in the world and it is native to South American; however, the American continent is the second largest producer with Brazil as its major producer. Pineapple fruit has high moisture content and the active metabolism increases the deterioration of the fruit soon after harvest. There are substantial losses of fruits in the field and in post-harvest, which are mainly caused by not using appropriate technology in cultivation, handling, storage and marketing. Due to its perishable nature chilling temperatures are widely used to extend the postharvest fruit quality during transportation. It is well known that temperature is usually the most important environmental factor affecting post-harvest life of fruit and vegetables; however, pineapples show chilling injury symptoms during storage at low temperature. For pineapple fruit, the optimal temperatures are 10°C and chilling may occur in the field in winter grown fruit causing pre-harvest chilling injury or after harvest when the fruit is stored at low temperatures. Minimal processing of fruits is an alternative technology to reduce post-harvest losses of perishable products and contribute to the further development of agribusiness in Brazil.

INTRODUCTION

Pineapple (*Ananas comosus* L. Merr.), a member of the *Bromeliaceae* family, (monocotyledons) has terminal inflorescences and a terminal multiple fruit. The pineapple plant only fruits once in its whole lifecycle, but slips and suckers produced from the mother stem are used as commercial vegetative propagation. It is a short tropical herbaceous perennial plant with a height of 1.0–2.0 m, native to Central and South America and is grown in several tropical countries such as Hawaii, India, Malaysia, the Philippines and Thailand. Overall there are about 2,000 species. Pineapple is one of the three most important tropical fruits in the world, after bananas and mangoes, and is also an important ornamental and textile plant. Its varieties are plentiful, but only a few leading types are sold commercially with an annual worldwide production of over 14 million tons. It is the eighth most produced fruit in the world (Chan et al. 2016; Elss et al. 2005; Espinosa et al. 2017; Hajar et al. 2012; Hazarika et al. 2017; Hong et al. 2013) and is consumed worldwide, both as fresh fruit and as industrialized products. Currently, the world production of pineapple is approximately 12.6 million tons. Thailand is the largest producer, accounting for 16% of world production, followed by the Philippines (12%) and Brazil (10%) where the "Pérola" variety is the most produced. The excellent qualitative characteristics of this fruit reflect in its socioeconomic importance. This excellent fruit was probably indigenous to Brazil and it had spread to other parts of tropical America. Nowadays, it is grown throughout the tropical and subtropical regions of the world and in Brazil this culture is economically exploited in most states. The Northeast (697,202 thousand fruits produced in 2014) and the Southeast (494.192 thousand fruits produced in 2014) are regions that have the largest cultivation. Nevertheless, almost all Brazilian states have excellent climatic conditions for its development and production, which makes it important to generate income, employment and help retain the rural population. In 2014 the pineapple was produced in all the federative units of Brazil, except the state of Piauí (IBGE, 2016). Pineapple was the first plant with artificially induced flowering on a commercial scale, which allowed this culture to be exploited economically. Artificial flowering of the pineapple meant that plants flourished at the same time, and crop harvests were concentrated which was economically an advantage. Also this practice enabled staggering of the harvests by plots as well as facilitating cultural and phytosanitary treatments. Pineapple is the most popular of the non-citrus tropical and subtropical fruits because of its attractive flavor and refreshing sugar-acid balance. The pineapple fruit is a good Data from of vitamin C to prevent oxidative damage in body cells by scavenging reactive oxygen species, as well as vitamin B to aid digestion, vitamin A, fibers and minerals. Pineapple fruit has been highlighted for its profitability and nutritional and organoleptic characteristics. These qualities are related to the size, shape, color, and the flavor of the fruit itself (Bengozi 2006; Hajar et al. 2012; Morais et al. 2014; Perecin et al. 2011; Yeoh and Ali 2017). Moreover, pineapples contains a high-activity protease called bromelain, a protein-digesting enzyme which has numerous biological promises and therapeutic applications, including prevention of tumor growth, blood coagulation, inflammatory changes, debridement of third degree burns and enhancement of drug absorption. This protein-digesting enzyme, found in the tissues of the *Bromeliaceae* plant family like pineapples is also used as a meat-tenderizer and is an important derivate from pineapple (Chakraborty, Rao, and Mishra 2016; Chaurasiya and Umesh Hebbar 2013; Liu et al. 2017; Matos et al. 2011).

PINEAPPLE

One of the most prized and popular fruits worldwide, historians believe that the pineapple is native to southern Brazil and Paraguay where the original wild plants grew and were spread by the local Indians up through South and Central America and on to the West Indies. The pineapple (*Ananas comosus* var. comosus) was probably first domesticated in the Guiana Shield, and *A. comosus* var. *ananassoides* is its wild ancestor. Later, when Christopher Columbus discovered Americas in 1493, he found the fruit, already domesticated, on the Guadeloupe islands and took it back to Spain. Columbus and his crew were probably the first Europeans to have tasted the fruit. It spread around the world on sailing ships that carried the fruit to protect the crews against scurvy, a disease caused by vitamin C deficiency. Particular features of pineapple associated with asexual reproduction, wide adaptation to different environments, and the evidence of the beginning of agriculture in the Amazon region in the late Pleistocene suggest that its domestication started between 6,000 and 10,000 years ago. This process led to ample genetic variability in numerous morphologically diverse cultivars (California Rare Fruit Growers 1996; Erin Cooks n.d.; Samson 1986; Scherer et al. 2015). The pineapple also known as Ananas, Nanas and Pina, is a monocotyledonous tropical or subtropical fruit of commercial importance belonging to the family *Bromeliaceae*, and genus *Ananas*. The *Bromeliaceae* family has 58 genera and more than 3,000 species and its members usually produce large amounts of proteases with no apparent function in plant growth and development. The pineapple plant is terrestrial and generally develops in the open field under high luminosity, sandy soils and tropical climate. It is small in size, measuring about 1.20m in height, its roots are poorly developed and the leaves, arranged spirally and grouped in the base form a rosette, are long and hard, linear in the form of a gutter and their margins are aculeated. Small flowers, pink to purple, appear agglomerated on a stem, forming the inflorescence that will develop originating the pineapple fruit, by the coalescence of individual berry type fruits. The fruit of the pineapple is a sorose, which means a collective fruit or a fleshy multiple fruit formed by the coalescence of the individual fruits, which merge on a central axis. The inflorescence develops from the base to the apex, and as a result, the base buds are physiologically older than those at the apex of the fruit. In addition, it is a non-climacteric fruit, and consequently, its eating quality is determined at the time of harvest, with little variation thereafter, despite the fact that the flesh fruit quality attributes vary along the fruit. The establishment of indices for harvesting mature fruits and the techniques for transport and processing are of great economic interest (Bengozi 2006; Chakraborty, Rao, and Mishra 2015; Diva et al. 2011; Montero-Calderón, Rojas-Graü, and Martín-Belloso 2010; Nadzirah et al. 2013; Soares, Trugo, Gonçalves, et al. 2005). The best temperature for the development of the roots and leaves of the pineapple range from 22°C to 32°C. However, because it is a tropical crop, the pineapple can tolerate higher temperatures, even slightly above 40°C. Although, relatively long periods of temperatures much higher than 40°C, associated with high insolation, can cause burning of leaves and fruits, resulting in production losses. The same is true for the night time temperatures, i.e., very low temperatures, below 5°C, have negative effects on the development and production of pineapple. The pineapple taxonomy was recently revised and simplified by Coppens d'Eeckenbrugge and Leal (2003) who downgraded the two genera: *Ananas* and *Pseudoananas* and the seven species of the Smith and Downs (1979) classification, to two species: *A. comosus* (2n = 50) and *Ananas macrodontes* (2n = 100). According to this new classification *A. comosus* is subdivided into five botanical varieties: var. *comosus*, var.

ananassoides, var. *erectifolius*, var. *parguazensis* and var. *bracteatus*, which include the former (Smith and Downs, 1979) diploid species. *A. comosus* var. *comosus* is the domesticated form with the largest fruits and the most widely commercialized pineapple. *A. comosus* var. *bracteatus* is cultivated as hedges and for fiber production, and used in traditional medicine. Brazil, Thailand, Philippines, and China are the main pineapple producers and are responsible for 50% of the world production. Although pineapple is considered a plant relatively undemanding in terms of water, regions with a total rainfall of between 1,000 and 1,500 mm per year are ideal for commercial cultivation, provided that the rainfall is well distributed. Curiously, pineapple is cultivated in regions of South Africa and Northeast Brazil, among others, where the precipitation varies from 500 to 600 mm as well as in regions of Costa Rica whose rainfall exceeds 3,000 mm (Liu et al. 2017; Matos et al. 2011).

Pineapple is a tropical fruit with attractive sensorial properties such as mouth feeling, flavor, good acidity/sweetness ratio, color and nutritional benefits providing several essential minerals such as calcium, phosphorus and iron, vitamins (B1, B2 and C), antioxidants and fiber. It is also low in calories, but rich in carbohydrates, is fat free, a good Data from of antioxidants because of high levels of phenolics and it is very versatile. It can be consumed raw, as juice, cooked, dried or canned and offers tremendous nutritional value (Offia-Olua and Ekwunife 2015; Rodríguez et al. 2016). This fruit has an exotic aroma, taste and flavor that are greatly appreciated. Also due to its proteolytic activity it acts as an adjunct in the digestion of food. Fully ripe fruits that exhibit the most intense flavor are generally only available in the producing countries; however they appear in a premium niche high-end segment of foreign markets, exported by fast-loading aircraft. Volatile pineapple profiles are related to fruit maturity; however, most studies on volatile pineapple fruit compounds do not provide sufficient information on the influence of harvest maturity, fruit logistics and post-harvest storage on these volatile compounds. Lactones are prominent as important contributors to the taste of fully mature pineapples (Morais et al. 2014; Steingass, Langen, et al. 2014) The Smooth Cayenne cultivar is the most popular in the world and about 70%

of the world production corresponds to this cultivar. Pineapples are highly perishable and seasonal; however, they can be consumed and served fresh, as juice and cooked. Mature fruit contains about 14% sugar and its sugar/acid balance contributes to its refreshing flavor. Fruit sweetness is a major factor determining its quality and sweetness gradually increases during the later stages of its growth. Variation in pineapple fruit sugar content is associated with fruit maturation and growing conditions. The composition of pineapple juice varies depending on geographical location, season, process and time of harvest. Sweetness depends on the concentration of sugar, which is synthesized and accumulated in the flesh during growth. There are indications that increasing the potassium level in the plant provides better flavor and aroma of the fruit, in addition to increasing the diameter of the stalk. Potassium also increases the ascorbic acid content that reduces quinones produced by enzymatic oxidation of phenols, converting them to hydrocoxic acid and acting as an inhibitor of the enzyme polyphenoloxidase (PPO), responsible for internal darkening of the pulp, known as internal browning (IB). Green pineapple is also used for making pickles and in various food items like squash, syrup, jelly, vinegar, alcohol, citric acid, calcium citrate. Approximately 60% of fresh pineapple is edible (Dorey et al. 2016; Farid Hossain, Akhtar, and Anwar 2015; Maeda et al. 2011; Soares, Trugo, Botrel, et al. 2005).

Production and Floral Induction

Pineapple production represents approximately 25% of the world tropical fruits. Although very profitable, it demands attention. The flowering process is uneven, which affects the regularity of production, resulting in fruits not included in standard commercialization.

Natural flowering of pineapple can occur under different environmental conditions, including low temperatures, short photoperiods, geotropic stimulation and other stress conditions. Once reproductive development is initiated, pineapple inflorescence and fruit development continue without interruption until fruit maturation. However, from an economical point of view, it is desirable that all the plants in the same field in a pineapple plantation flower simultaneously and this can be achieved with artificial floral induction. There are many chemicals that induce inflorescence in pineapple that are used commercially such as the sodium salt of α-naphthaleneacetic acid (SNAA), ethylene and ethephon and there is calcium carbide, which is widely used in Brazil by small producers. (Figure 1) Exogenous ethylene, or ethephon, has been widely used to induce pineapple flowering, but the molecular mechanism behind ethephon induction is still unclear (Bartholomew 1977; Carvalho et al. 2005; Y. Li et al. 2016).

(a) (b)

(c) (d)

Figure 1. Flowering of pineapple Cv Turiaçu (A). Application of calcium carbide, 0.5 to 1g per plant (B). Manual application of ethephon with manual carry back pump (C), Mechanized application of ethephon with the use of a automatic sprayer (D). Photo: Araujo, J.R.G.

Factors such as temperature, rainfall, photoperiod, luminosity, wind and relative humidity play an important role in the development of pineapple. The photoperiod has a direct influence on flowering, since, in periods of shorter days there is a higher occurrence of early natural flowering. Also luminosity influences the development of the plant, the production and the quality of the fruit, with the optimum luminosity ranging between 2,500 and 3,000 h/year. A 20% reduction in brightness results in a 10% reduction in yield. The wind, more specifically strong and dry winds, besides having their direct effect on the aerial part, as they are able to tilt the plant, can also cause negative effects, especially in the fruits, such as cracks due to low relative humidity. Table 1 shows the climatic characteristics of some pineapple producing regions of the world.

Table 1. Climatic characteristics of some pineapple producing regions around the world. Data from: (Matos et al. 2011)

Location	Temperature (°C)			Average rainfall (mm)
	Max	Min	Average	
Johore (Malaysia)	35.0	18.9	26.9	2,880
Nyombe (Cameroon)	-	-	-	3,000
Buenos Aires (Costa Rica)	32.0	19.0	23.0	3,078
Arecibo (Puerto Rico)	-	-	25.3	2,190
Wahiawa (Hawaii)	20.0	14.3	22.6	1,062
Assam (India)	34.0	7.0	-	2,300 to 3,800
Thika (Kenya)	35.0	5.5	20.5	803
Coração de Maria (Brazil)	30.0	17.3	23.6	1,150
Itapemirim (Brazil)	36.0	16.0	26.0	1,162
Rock Hampton (Australia)	27.2	16.7	22.7	1,002
Port Elizabeth (South Africa)	21.2	13.3	17.2	577
Itaberaba (Brazil)	30.8	19.3	24.6	762

Correct management and care during pre-harvest of the pineapple has a direct influence on the quality of the fruit harvested, the shelf life and post-harvest quality. The cultural treatments of pineapple involve the management and control of invasive plants, integrated control of fusariosis, floral induction with phytoregulators, and protection of fruits against solar burning, among others. The spacings used in pineapple plantations vary widely according to the cultivar, destination of the production, level of mechanization and other factors. The main planting systems adopted in single rows are: 0.90m x 0.30m or 0.80m x 0.30m, corresponding to 37,030 and 41,660 plants/ha, respectively, and double rows, 0.90m x 0.40m x 0.40m or 0.90m x 0.40m x 0.35m or 0.90m x 0.40m x 0.30m, resulting in densities of 38,460, 43,950 and 51,280 plants/ha, respectively. Double-row planting should be associated with the use of herbicides to control weeds and planting in this mode should be alternated, that is, plants in a row are planted in line with the empty spaces of the other row (GANEM, 2015).

Control of Invasive Plants

The control of invasive plants can be carried out by manual weeding, chemical weeding (herbicides), plant cover (using dead cover or synthetic material such as plastic) and use of mechanical equipment such as manual brush-cutters (Figure 2). The competition caused by weeds is more damaging in the first few months after planting, which corresponds to the rooting period and slow growth of the pineapple. In the reproductive phase, this is, after the floral induction of pineapple, competition is minimal and practically does not affect the fruit production or weight, and control of invasive plants is unnecessary (Ganem 2015; Sanches and Matos 2013). A consortium of crops is one of the practices that can be used to reduce invasive plants. This practice is viable in pineapple production as it reduces implantation costs, provides extra income and controls invasive plants. Pineapple can be intercropped with beans, peanuts, sunflowers, watermelons, okra, cabbage, tomatoes and other short cycle crops, which are planted between the lines and at the same time as the pineapple crop. The consortium should be restricted to the first six months of the pineapple cycle, avoiding the consortium of pineapple with maize, which is host to gummy or fusariosis. Pineapple is also used in consortium with other longer cycle fruits such as in citrus fruits, guarana, coconut, guava, graviola, and mango, among others; however, in this case, pineapple is the secondary culture (Ganem 2015; Silva 2007).

Figure 2. Control of invasive plants by: A - Dead cover with millet in pre-planting, Photo: Matos, A.P. (2011); B - Use of plastic films on flowerbeds with equipment developed by Agrobrás in Bauru (SP), Photo: Sampaio, A.C. (2011); C - Field test with plastic film, in Guaraçai (SP), Photo: Suzuki, R.T. (2012); D - Use of brushcutters in certified area in Integrated Production in Tocantins, Photo: Matos, A.P. (2011); E - Application of herbicide using the Motoagro equipment (Coupled Kit of Pruning, Fertilization and Spraying); F - Motoagro spraying machine for spraying pesticides and herbicides. The control of invasive plants can be accomplished by manual weeding, chemical weeding (herbicides), vegetal cover through dead cover or synthetic material (plastic) and use of mechanical equipment such as manual brushcutters (Ganem 2015; Motoagro 2015).

Crop rotation, on the other hand, helps to break the weed cycle, preventing the production of more seeds and avoiding an increase in the weed seed bank, which is an important practice for weed control. Failure to use this practice can lead to edaphic and nutritional problems in some producing regions after years of intensive pineapple cultivation. The use of soil cover with organic materials provides enormous agricultural benefits and the adequate choice of cover is fundamental to achieve the proposed objectives. Mulching is a good alternative, as it helps to control the emergence of weeds, maintains soil moisture, controls erosion and adds organic matter to the soil. On the other hand, planting in black plastic film mechanically fixed in the soil (mulching system) is also a good option for the control of invasive plants (Cunha 2004; Ganem 2015; Meletti, Sampaio, and Ruggiero 2011; Sampaio, R. A.; Araújo 2001). The plastic covers the entire planting area, not allowing weeds to grow, with holes being opened only where the seedlings are planted. The manual, gasoline brushcutter (25cc) to keep weeds under control in between planting lines is also a good alternative. This is efficient in the management of weeds, presents environmental, economic and social advantages. In environmental terms, the aerial part of the invasive plants remains on the ground, and becomes mulch thus helping to reduce the chemical control (herbicide application). From a social point of view, the gasoline brushcutter of invasive plants associated with the pineapple crop is an activity that requires less effort compared to manual weeding and compared with chemical control does not present any risks to human health, since the worker is not exposed to the negative effects of herbicides. In economic terms, one hectare can be cut by one man in a day with an average fuel consumption of three liters. Manual weeding of an equal area takes 10 men one day, which is seven times the former cost. The alternative is to control the weeds with herbicides, especially in large and heavily planted plantations. However, the application must be carried out with care so that the pineapple does not suffer from the possible toxic effects of the products applied. To avoid this problem it is essential to calibrate the sprayer before application. It should be emphasized that the herbicides indicated for pineapple cultivation are phytotoxic to other crops such as beans, which makes it impossible to use them in a consortium (Cunha 2006; Ganem 2015).

DISEASES

The postharvest losses of pineapple fruit remains a substantial problem for producing countries. The major world pineapple producers are Costa Rica, Brazil, Philippines, Thailand, and Indonesia. However, even Brazil that is one of the largest producers, its yield of 41 t. ha^{-1} with an average of 2.478 million tons in 61,000 ha is far lower than that of Indonesia, which is 124.5 t ha^{-1}. Some factors contribute to these low yields such as adverse environmental factors, inadequate cultural practices and phytosanitary problems. The occurrence of diseases in plants limits high yields and may even make production unviable. Diseases, weeds and pests markedly affect the development of the pineapple crop, requiring the use of pesticides. There are several diseases that affect the pineapple crop and about 85 different species of organisms have been found only in this culture. Fusariosis caused by *Fuzarium guttifotme Nirenber; O'Dennell* requires special attention and is the main limiting factor for pineapple cultivation in Brazil and other South American countries. This pathogen was first reported infecting pineapple fruits of the cultivar Smooth Cayenne in 1964 in Bolivia. Considering that the majority of the main pineapple cultivars such as MD-2, Pérola, Queen, Red Spanish, Smooth Cayenne, among others, are susceptible to *Fuzarium sp.*, this pathogen becomes even more important (Garcia et al. 2017; Júnior 2009; Nogueira et al. 2014; De Souza, Trocoli, and Monteiro 2016).

*Fuzarium sp. is a*ble to infect the whole plant, including the fruits. This pathogen infects the seedlings at an early stage of development, and may cause losses as high as 80% depending on the harvest season, region of production, and inoculum potential. Infection of vegetative planting material may be as high as 40% and around 20% of the infected planting material that is brought to the field dies before treatment to force flowering.

One of the most evident symptoms is gum exudation, which can manifest in all stages of plant development, especially in the fruits (Figure 3). It is barely perceptible in the early stages of the disease; however, as the fungus survives in seedlings and plants, knowledge of the symptoms means that elimination is an important Data from of inoculum. But the ability of the pathogen to infect the vegetative planting materials makes the disease even more important since pineapple is a vegetatively propagated crop. Traditionally pineapple fusariosis is controlled by preventive fungicide applications during inflorescence, from open heart to the dry petal stages. However, treatment of infected seedlings with fungicides prior to planting should be carefully evaluated because it only eliminates the external inoculum, that is, it does not have a curative action even at high doses and regardless of the time of treatment. (Carvalho et al. 2003; Gomes et al. 2003; De Souza et al. 2016).

The control of fusariosis requires an integration of several cultural practices such as the use of healthy propagative material, constant inspection at planting, removing the infected material and carrying out chemical control (Garcia et al. 2017; Khayankarn et al. 2013; Morais et al. 2014; Selvarajah, Bauchot, and John 2001; Silva, Fernandes, and Mauro 2014; Soares, Trugo, Botrel, et al. 2005).

Resistant cultivars are considered a more sustainable management option. Cultivars resistant to fusariosis and with desirable agronomic characteristics have been obtained through genetic improvement, thus constituting the most efficient and economical method of control. Recently the cultivar Vitoria, resistant to fusariosis, has been released to growers and has attracted interest in Brazil. Also fusarium resistant hybrids with smooth leaves have been obtained recently; cultivars such as: BRS Vitória, BRS Imperial and IAC Fantástico. However, they have not yet been adopted throughout the country and so there is lack of information about the stability of their resistance to the pathogen in the long term. Remembering that the most common cultivar in the world (Smooth Cayenne) and the most commonly cultivar planted in Brazil (Perola) are both susceptible to fusariosis (Aquije et al. 2010; Cabral and Matos 2009;

Ferreira, F.R.; Cabral 1998; Ganem 2015; De Souza et al. 2016; Spironello et al. 2011; Ventura et al. 2009).

(a) (b) (c)

Figure 3. Fusariosis in pineapple: gummy exudation in molt (A); Gumous exudation on fruit (B); Mummified fruit in the final stage (C) Data from: (Carvalho et al. 2003).

The cultivar 'Turiaçu', native to the municipality of Turiaçu - MA, Brazil, selected by family farmers, behaves under field conditions as resistant to fusariosis. It presents fruits that are highly appreciated in the regional consumer market, due to the high sweetness, which is higher than the cultivar Perola, with medium mass and yellow flesh. The cultivar is in the process of domestication, and already possesses clonal selections that will soon be recommended to producers. Currently, these clonal selections are being tested for resistance to fusariosis and their technological aptitude by researchers from the State University of Maranhão (UEMA) and Embrapa Food and Technology (Araujo et al. 2012).

Extensive application of chemical pesticides is the normal control practice in commercial plantations, but this has become highly questionable due to elevated costs, deteriorating soil health and general environmental issues. The demand for the systematic use of fungicides, in addition to increasing costs, can still affect resistant strains of the pathogen, when used inappropriately, making it even more difficult to manage the disease. An alternative to the use of pesticides in disease control is the induction of resistance, which activates latent resistance mechanisms of the plant with the use of biotic or abiotic agents (Aquije et al. 2010; Uchôa et al. 2014).

POST-HARVEST

Pineapple has a short shelf life, which increases postharvest losses due its high moisture content. In addition, its active metabolism increases the deterioration of the fruit soon after its harvest. The maturation stage at which the fruit is harvested is critical to its shelf life as well as to its storage potential. The characterization of the harvest point is very important because if it is carried out before the product has completed its physiological development then the ripening process is compromised. In the same way, if the harvest is carried out when the product is very mature there will an increase in losses. Also pineapple has a perishable nature, especially when exposed to chilling temperatures. Under conditions of low temperatures the fruits rapidly develop internal browning (IB), which severely restricts exports using refrigerated sea freight. IB, which is also known as endogenous brown spot or blackheart, has caused heavy losses in pineapple production and is the most important physiological disorder when pineapples are stored at temperatures below 13°C post-harvest and it develops quickly within 2–3 days when the fruits are moved to temperatures of 15-30°C. The results are severe internal discoloration of pineapple fruit. (Hong et al. 2013; A.-K. Raimbault et al, 2013; Selvarajah et al. 2001, Soares, Trugo, Gonçalves, et al. 2005; Steingass, Grauwet, and Carle 2014).

IB is responsible for dampening the export potential of this fruit from the producing countries. It limits both the storage and the export of this fruit and so it is very important to find methods of extending the shelf life and maintaining adequate fruit quality, especially reducing the occurrence of IB, the most important injury to pineapple. The final quality of pineapple, however, depends on many factors from planting to harvesting technologies and postharvest procedures. Beside some varietal and harvest-maturity differences, fruit quality is strongly affected by post-harvest handling. Chilling injury as we have seen is very important and is a complex physiological disorder that occurs in tropical and sub-tropical fruit, including pineapple, after exposure to low temperatures (0-20°C). (Lu et al. 2011; Luengwilai, Beckles, and Siriphanich 2016; Steingass, Grauwet, et al. 2014).

Heat, waxing, atmosphere control post-harvest salicylic acid treatment and the application of 1-methylcyclopropene are postharvest treatments that individually are inefficient to prevent internal IB. However, the combination of these preservation technologies with low temperature storage, which is well known the most important environmental factor affecting post-harvest life of fruit and vegetables, can exert a beneficial effect on maintaining the quality of pineapple fruits than when applied alone. Transgenic strategies to control the browning process in other crops, such as potato, are interesting. A similar strategy, based on silencing the PPO gene, is being developed for pineapple and potentially will be able to control IB injury in pineapple fruit. However, the efficient exploitation of this biotechnology to control IB relies on a precise understanding of the biochemical pathway in IB development. Besides IB being a physiological disorder of pineapple that is induced by exposure to low temperatures, either in the field or in postharvest storage, the biochemical pathway during low temperature storage has not been clearly documented. PPO and phenylalanine ammonia lyase (PAL) have been shown to play important roles in the browning process of many fruits and vegetables. On the other hand, evidence has increased to suggest that chilling injury in other species is mediated by the balance of reactive oxygen species production and the cell protective system during low temperature stress or upon release from the stress (HONG et al., 2013; SOARES et al., 2005; ZHOU et al., 2003a).

Pineapple is a good Data from of ascorbic acid, thus it is essential to avoid the loss of this vitamin, which commonly occurs during cold storage and poor marketing conditions. Adequate mineral plant nutrition is crucial for good quality and productivity and the appropriate concentration of potassium (K) in the soil appears to play a relevant role in producing good quality pineapple promoting an increase in total solids and producing fruit with larger diameters. However if the K levels are too high, the fruits may become very acidic with pale and rigid pulps. There are indications that appropriate K levels in the soil may also contribute to improved pineapple flavor, to increased stalk diameter and to increased levels of ascorbic acid. The latter may prevent some degree of enzymatic browning by inhibition of PPO activity. (Lu et al. 2011; Selvarajah et al. 2001; Soares, Trugo, Botrel, et al. 2005; Zhou et al. 2003). Among the enzymes present in pineapple, undoubtedly bromelain is one of the most important. Bromelain is a proteolytic digestive enzyme that when taken with meals, helps the digestion of proteins, working to break proteins down into amino acids. Brazil still imports bromelain from Germany and Switzerland but this enzyme could be extracted from the by-product of the juice processing industry or from the stalk of field crops after harvesting the seedlings (Farid Hossain et al. 2015; Meletti et al. 2011).

Bromelain

The global industrial enzymes market worth 3.3 billion and 4.2 billion US dollars in 2010 and 2014, respectively, continues to grow. Modern technological advancements have increased the protease production. Today, proteases dominate with approximately 60% of the market share for the worldwide enzyme market. However, the main challenge in the comprehensive application of enzymes in industry has been their high costs. Cysteine proteases comprise a family of enzymes widely distributed in nature and several have been isolated from a number of plant Data froms. One of them is bromelain, a mixture of cysteine proteases with high activity that is extracted from the pineapple plant. It is the main protease enzyme present in the pineapple and constitutes an unusually complex mixture of protein digesting enzymes including thiol proteases and non-protease components. The proteases are the main components of bromelain and include stem bromelain (SBM) (80%), fruit bromelain (FBM) (10%), and ananain (5%). Among the non-protease components are phosphatases, glucosidases, peroxidases, cellulases, glycoproteins and carbohydrates. Its presence in pineapple fruit was first assigned the EC number (EC 3.4.22.4), later the stem bromelain was assigned (EC 3.4.22.32) and fruit bromelain (EC 3.4.22.33) a collective name for proteolytic enzymes or proteases found in tissues including stem and fruit. However, bromelain can also be isolated in small amounts from pineapple waste such as the core, leaves, and peel. Stem bromelain (SBM) is different from fruit bromelain (FBM). Although there are different EC numbers, both the fruit and stem bromelain isoforms are founded in pineapple fruits. The bromelain concentration is higher in the stems than in the fruit, making it a cheaply available Data from of bromelain (Arshad et al. 2014; Hazarika et al. 2017; R. Li et al. 2016; Manzoor et al. 2016; Mohan et al. 2016; Nadzirah et al. 2013; Pavan et al. 2012; A. K. Raimbault et al. 2013).

Cysteine proteases in plants have been shown to be involved in development, protein turnover, degradation of damaged proteins and in response to various abiotic and biotic stresses but until today, the biological role of bromelain in pineapple remains unknown. This enzyme has been known chemically since 1876 and was identified for the first time by Marcano in 1891 but the isolation of enzymes from pineapple only started in 1894 when the extraction and purification of bromelain from the natural pineapple fruit first began to be investigated, as it still is today. Sulfhydryl proteolytic enzymes are the chief constituents of bromelain. A single polypeptide chain constitutes the primary structure of SBM with 212 amino acids and the molecular weight is 33 kDa folded into two structural domains stabilized by disulfide bridges and numerous hydrogen bonds. The active site is located on the surface molecules between domains, with two catalytic residues, Cys25 and His159, for hydrolysis of cleaved bonds and substrate specificity. The first N-terminal domain of SBM contains mainly β- sheets while the C-terminal domain is composed of α-helices, and therefore it is classified within the α+β protein class, which includes other cysteine proteases like papain, actinidin and chymopapain.

This proteolytic enzyme has unique functions useful for the food, pharmaceutical and cosmetic industries that make it one of the best vegetal proteases with a high commercial value, costing up to US$ 2,400 per kg. When taken with meals, bromelain aids in the digestion of proteins, breaking them down into amino acids, it has anti-inflammatory properties on an empty stomach. Moreover, sinusitis, burns, pancreatic insufficiency and skin rashes seem to benefit from the ingestion of bromelain, according to the national library of medicine. Bromelain also has a potential as an anti-cancer agent as it affects major pathways and regulators implicated in cancer. Bromelain has also been envisaged to have extensive applications as an active ingredient in tooth-whitening dentifrices and skin products. Table 2 show the main applications of bromelain (Arshad et al. 2014; Devakate et al. 2009; Manzoor et al. 2016; Nadzirah et al. 2013; Soares et al. 2012; Vilanova Neta et al. 2012). However, bromelain proteases are usually unstable and sensitive under stress conditions, such as in the presence of elevated temperatures, organic solvents and chemicals, which may result in a decrease in the enzymatic activity of this health-promoting enzyme, limiting its pharmacological, industrial and biotechnological applications. Unlike crude SBM, which is used widely in industry, FBM is not commercially available despite the large quantities of waste pineapple fruit discarded at pineapple canneries. The production of bromelain from pineapple is facing great challenges to reduce the extraction costs. In industry, ultrafiltration followed by precipitation and freeze drying is used for the large- scale separation of bromelain.

However, researchers have conducted many investigations on ways to isolate this enzyme from the crude pineapple extract to produce a pure bromelain in less steps and lower costs. Common techniques for the purification of bromelain from the crude pineapple extract include reverse micellar system, two-phase aqueous extraction, cation exchange chromatography and ammonium sulfate precipitation. SBM and FBM are both single-chain glycosylated enzymes but they possess different characteristics. SBM has reduced proteolytic activity and exhibits lower specificity for peptide bonds then FBM (Devakate et al. 2009; R. Li et al. 2016; Manzoor et al. 2016; Nor et al. 2016; Offia-Olua and Ekwunife 2015; Pavan et al. 2012).

Table 2. Industrial applications of bromelain.
Data from: (Arshad et al. 2014)

Applications	Reasons
Baking industry	-Improve dough relaxation and allow the dough to rise evenly; -Produce hypoallergenic flour.
Tenderization	-Hydrolyze meat myofibril proteins; -Hydrolyzing agent for meat, oyster, chicken, and squid.
Fish protein hydrolysate	-Hydrolyze fish protein to generate fish protein hydrolysate.
Anti-browning agent	-Inhibit browning of fruits and phenol oxidation.
Alcohol production	-Enhance protein stability of beers Prevent haze formation.
Animal feed	-Estimate protein degradation in ruminant feed.
Textile industry	-Minimize softening time in cocoon cooking; -Remove scale and impurities of wool and silk fibers; -Enhance dyeing properties of protein fibers.
Applications	Reasons
Tooth whitening	-Remove stains, plaque, and food debris on the outer surface of teeth.
Cosmetic industry	-Treat acne, wrinkles, and dry skin; -Reduce post-injection bruising and swelling.

The body can absorb significant amounts of bromelain; about 12 gm/day of bromelain can be consumed without any major side effects. This enzyme influences blood coagulation by increasing the serum fibrinolytic ability and by inhibiting the synthesis of fibrin, a protein involved in blood clotting. Moreover, bromelain has been shown to be an anti-inflammatory agent and has been used in medical practice for acute inflammation and several other related conditions. The anti-inflammatory action of bromelain is mediated by the retardation of pro-inflammatory prostaglandin formation. Bromelain is still an effective mucolytic agent and used efficiently in rhinitis, rhinosinusitis as well as in chronic rhinosinusitis (Manzoor et al. 2016; Pavan et al. 2012).

VOLATILE COMPOUNDS

Eating is an activity that integrates all the sensory modalities. Taste is activated when the chemical stimuli reach the taste buds in the mouth. Such an environment involves saliva, retronasal smell and tactile information. Flavor results from the combination of these three modalities: taste, smell, and tactile processes. Aromatic substances have a high sensory/physiological value and influence the appetite and digestive system significantly; they can even affect the mental state of people. This olfactory perception is triggered by the release of volatile compounds (VC) present in fruits and the consumption of tropical fruits is mostly related to their unique and exotic flavor rather than their nutritional value. Flavor is the result of the interaction of chemical constituents with the consumer's sense of taste and smell and it is composed of volatile and non-volatile compounds that cause taste sensations; this is referred to taste and odor perception. Aroma can objectively reflect the characteristics of different fruits, and is certainly considered one of the most important indices of fruit quality. Consequently, much attention has been given to the study of volatiles in fruits, as well as for synthesizing the essence and for picking, storage and cultivation. VC are primarily responsible for the single flavor characteristics that distinguish different fruits and vegetables and determine their desirability to the consumer. These substances are produced through metabolic processes during field maturation, the post-harvesting period and even in the storage of the same (Bugaud and Alter 2016; Dellacassa et al. 2017; Razumiejczyk et al. 2016; Soares, Trugo, Gonçalves, et al. 2005; Steingass, Grauwet, et al. 2014; Ye et al. 2017; Zhang et al. 2013). Pineapple is an exotic fruit that is highly appreciated for its aroma, juiciness and flavor (taste and aroma). There are many cultivars, with different colors, shapes, sizes, odor, and flavors. One of them, the Gold cultivar (MD2) has stood out in the international markets because of its sensory characteristics, highlighting flavor, sweetness to acidity balance and juiciness. The VC of pineapple have been studied for over 60 years and more than 280 compounds are known to be involved in generating the characteristic pineapple flavor. Sweetness, sourness and aroma are the main components of fruit flavor, given by the balance among sugars, acids, and volatiles. In fact, the perception of sweetness can be modified by the acid content and aroma compounds (Montero-Calderón et al. 2010; Wei et al. 2011; Zheng et al. 2012). (Montero-Calderón et al. 2010). VC are produced through metabolic pathways during harvest, postharvest and storage and depend on many factors related to the species, variety, different geographical areas where the pineapple crop is grown, different seasons, different stages of ripening, development of the fruit and postharvest storage conditions. Fruit maturation involves a number of complex biochemical reactions, such as starch hydrolysis, chlorophyll transformations, carotenoid, anthocyanins and phenolic production and the formation of volatile substances. All these reactions are important for the final characteristics of the mature fruit and the formation of the specific flavor and aroma of each fruit. However, knowledge about the impact of harvest maturity and post-harvest operations, in particular, fruit logistics on biogenesis of volatile composition of fresh pineapples, is limited. Glycosidically bound volatile compounds are nonvolatile aroma precursors in many fruits, such as pineapple. The VC found in pineapple include a variety of lactones, acids, hydrocarbons, sulfur-containing compounds, carbonyl compounds and esters. Esters such as 2-methylbutanoates and hexanoates give fruity notes to fresh pineapple as well as other fruits. 4-Hydroxy-2,5-dimethyl-3(2H)- furanone (HDF; furaneol®; pineapple furanone), which possesses a pineapple and caramel-like aroma, was first isolated by Rodin et al. from pineapple. Despite the fact that more than 280 volatile compounds have been identified among the aroma volatiles of pineapple, only a few of these volatile compounds have been identified as contributors to the pineapple aroma, Table 3 (Kaewtathip and Charoenrein 2012; Ren et al. 2015; Rodin et al. 1965; Soares, Trugo, Gonçalves, et al. 2005; Steingass, Grauwet, et al. 2014; Tokitomo et al. 2005; Zheng et al. 2012).

Table 3. The volatile aroma compounds of two pineapple varieties. Data from: (Zheng et al. 2012)

Classification	Compounds Name	Content (μg.Kg^{-1})	
		Tainong n° 6	Tainong n° 4
Esters	Butanoic acid, 2-methyl-, methyl ester	19,48	-
	Butanoic acid, 2-hidroxy-2- methyl-, methyl ester	-	32.27
	Butanoic acid, 2-methyl-, ethyl ester	22.24	-
	Hexanoic acid, methyl ester	4.71	67.75
	Hexanoic acid, ethyl ester	8.35	-
	2,4-Hexadienoic acid, methyl ester	1.37	-
	3-(methylthio) propanoic acid methyl ester	32.94	622.49
	Butane-2,3-diyl diacetate	12.23	-
	3-(methylthio) propanoic acid ethyl ester	78.06	32.96
	Octanoic acid, methyl ester	20.52	142.25
	Ethyl 3-hydroxyhexanoate	8.61	-
	Benzoic acid, ethyl ester	9.68	-
	Octanoic acid, ethyl ester	46.21	-
	Benzeneacetic acid, ethyl ester	0.73	-
	Dodecanoic acid, methyl ester	18.28	16.33
	Methyl cinnamate	6.54	
	Decanoic acid, ethyl ester	19.96	
	Ethyl cinnamate	15.38	
	Dodecanoic acid, ethyl ester	0.97	
Total		326.26	917.05
Terpenes	(±)-Dictyopterene A	15.61	-
	4,9-Muuroladiene	3.64	-
	α-Muurolene	3.60	12.03
	(-)-Alloaromadendrene	1.26	-
	1,6-Cyclodecadiene	0.52	-
Total		24.63	12.03

Classification	Compounds Name	Content (μg.Kg^{-1})	
		Tainong n° 6	Tainong n° 4
Lactones	γ-Octalactone	12.49	-
	5-Hydroxyoctanoic acid lactone	9.93	-
	δ-Octalactone	-	63.40
Ketones	2,5-Dimethyl-4-hydroxy-3 (2H)-furanone	-	76.47
Total		-	76.47
Alcohol	2,3-Dimethyl-undec-1-en-3-ol	-	6.25
Total		-	6.25
Aldehydes	Decanal	1.61	-
Total		1.61	-
Acids	Octanoic Acid	-	8.24
	Ethyl (±)-3-acetoxybutyrate	3.90	
Total		3.90	8.24
Alkenes	3,4-Dimethoxystyrene	1.84	-
Total		1.84	-
Grand Total		**380.66**	**1080.44**

GASTRONOMY

Due to its attractive sweet flavor, pineapple is widely consumed as a fresh and canned fruit, as well as in processed juices and as an ingredient in exotic foods. Pineapple has strong perfume and varied flavor, sometimes very sweet and other quite acidic, besides refreshing pulp and full of broth. Such virtues recommend it as a fruit that lends itself to the production of jams, crystallized candies, jellies, juices, ice creams, creams and puddings besides, it can easily be processed into alcoholic and nonalcoholic drinks. Pineapples contain a good proportion of sugar and acids, which makes it suitable for winemaking. Amongst the various factors that contribute to enjoying these pleasant fermented beverages, the flavor characteristics are possibly the most important. Industry also produces different pineapple products (such as the minimally processed fruit and pineapple chips) aiming to facilitate consumption of the fruit and reduce losses. As a ready-to-eat product, fresh-cut pineapple is attractive to consumers and sold at many supermarkets and food distribution outlets in different shapes such as cubes, slices, chunks, and cored whole fruit. Fresh-cut pineapple has become more and more popular as it is considered to be convenient to use compared to whole pineapple (Dellacassa et al. 2017; Jiang et al. 2016; Liu et al. 2011; Silva et al. 2014; Soares, Trugo, Gonçalves, et al. 2005; Wei et al. 2011; Zhang et al. 2013). Pineapple is used in gastronomy as meat tenderizers due to its bromelain content, a protein-digesting enzyme extracted from pineapple fruit or its stem. Tenderness is an important characteristic of meat and the overwhelming demand for guaranteed tender meat has attracted players in the meat industry to provide an acceptable product of quality. The scientific processes in cooking were first introduced to the public in 1969 when Nicholas Kurti held a televised presentation "The Physicist in the Kitchen" for the Royal Society. In the presentation he explained the principles of microwave heating and demonstrated how pineapple juice tenderized pork (Arshad et al. 2014; Harold McGee 1987; Vartiainen, Aksela, and Hopia 2013).

Bromelain is also used in the baking industry. Gluten is a functional component of wheat food products, such as flour and consists of two major proteins, gliadin and glutenin. When hydrated, gluten becomes insoluble and forms lattice-like structures, so gluten must be degraded to avoid resistance to dough stretching. Bromelain improves dough relaxation, enhances solubility and even prevents dough shrinkage. This will allow the dough to rise evenly during the baking process. Moreover, bromelain also has been used to produce hypoallergenic flour that is suitable for wheat-allergic patients (Arshad et al. 2014).

References

Aquije, Glória Maria de Farias Viégas et al. 2010. "Cell Wall Alterations in the Leaves of Fusariosis-Resistant and Susceptible Pineapple Cultivars." *Plant Cell Reports* 29(10):1109–17.

Araujo, José Ribamar Gusmão, Rozalino Antonio Aguiar Júnior, Afonso Manoel Silva Chaves, and Fabrício De Oliveira Reis. 2012. "Turiaçu: A Pineapple Cultivar Traditional and Native from Maranhão, Brazil." *Rev. Bras. Frutic.* 34(4):1270–76.

Arshad, Zatul Iffah Mohd et al. 2014. "Bromelain: An Overview of Industrial Application and Purification Strategies." *Applied Microbiology and Biotechnology* 98(17):7283–97.

Bartholomew, D. P. 1977. "Inflorescence Development of Pineapple (Ananas Comosus [L.] Merr.) *Induced to Flower with Ethephon.*" 138(3):312–20.

Bengozi, Fábio José. 2006. "Origin, Seasonality and Phisical-Chemical Quality of the Pineapple Fruits Commercialized at the CEAGESP – São Paulo."

Bugaud, Christophe and Pascaline Alter. 2016. "Volatile and Non-Volatile Compounds as Odour and Aroma Predictors in Dessert Banana (Musa Spp.)." *Postharvest Biology and Technology* 112:14–23.

Cabral, J. R. S. and A. P. Matos. 2009. "Imperial, a New Pineapple Cultivar Resistant to Fusariosis." *Acta Horticulturae* 822:47–50.

California Rare Fruit Growers, Inc. 1996. "Pineapple Fruit Facts." *Copyright.* Retrieved January 27, 2017 (https://www.crfg.org/pubs/ff/pineapple.html).

Carvalho, Rêmulo Araújo et al. 2003. "Controle Agroecológico Da Fusariose Do Abacaxi Com Plantas Antibióticas." *EMEPA*. Retrieved February 14, 2017 (http://www.infobibos.com/Artigos/2006_2/abacaxi/Index.htm).

Carvalho, Sergio Luiz Colucci de, Carmen Silvia Vieira Janeiro Neves, Rodrigo Bürkle, and Celso Jamil Marur. 2005. "Floral Induction Period and Thermal Time Requirements from the Flowering to the Harvest Period for Smooth Cayenne Pineapple." *Revista Brasileira de Fruticultura* 24(3):430–33.

Chakraborty, Snehasis, Pavuluri Srinivasa Rao, and Hari Niwas Mishra. 2015. "Empirical Model Based on Weibull Distribution Describing the Destruction Kinetics of Natural Microbiota in Pineapple (Ananas

Comosus L.) Puree during High-Pressure Processing." *International Journal of Food Microbiology* 211:117–27.

Chakraborty, Snehasis, Pavuluri Srinivasa Rao, and Hari Niwas Mishra. 2016. "Modeling the Inactivation Kinetics of Fruit Bromelain in Pineapple during High-Pressure and Thermal Treatments." *Innovative Food Science & Emerging Technologies* 33:10–18.

Chan, Siew Ling, Yen Ping Tan, Abdul Halim Abdullah, and Siew Teng Ong. 2016. "Equilibrium, Kinetic and Thermodynamic Studies of a New Potential Biosorbent for the Removal of Basic Blue 3 and Congo Red Dyes: Pineapple (Ananas Comosus) Plant Stem." *Journal of the Taiwan Institute of Chemical Engineers* 61:306–15.

Chaurasiya, Ram Saran and H. Umesh Hebbar. 2013. "Extraction of Bromelain from Pineapple Core and Purification by RME and Precipitation Methods." *Separation and Purification Technology* 111:90–97.

Cunha, Getúlio Augusto Pinto da. 2004. *Cultivo Do Abacaxizeiro - Consorciação E Rotação de Culturas*. Cruz das Almas-BA. [*Pineapple Growing - Consortium And Crop Rotation*. Cross of the Souls-BA]

Cunha, Getúlio Augusto Pinto da. 2006. "Abacaxi on Line." *Regras Para Análise de Sementes* ["Pineapple on Line." *Rules for Seed Analysis*]. 4(1):3.

Dellacassa, Eduardo et al. 2017. "Pineapple (Ananas Comosus L. Merr.) Wine Production in Angola: Characterisation of Volatile Aroma Compounds and Yeast Native Flora." *International Journal of Food Microbiology* 241:161–67.

Devakate, R. V., V. V. Patil, S. S. Waje, and B. N. Thorat. 2009. "Purification and Drying of Bromelain." *Separation and Purification Technology* 64(3):259–64.

Diva, Correia, Neiliane Santiago Sombra Borges, Esaú Matos Ribeiro, and João Paulo Saraiva de Morais. 2011. *Produção de Mudas in Vitro de Abacaxi Ornamental*. Fortaleza, CE [*Production of seedlings in vitro of Ornamental Pineapple*. Fortaleza, CE].

Dorey, Elodie, Patrick Fournier, Mathieu Léchaudel, and Philippe Tixier. 2016. "Modeling Sugar Content of Pineapple under Agro-Climatic

Conditions on Reunion Island." *European Journal of Agronomy* 73:64–72.

Elss, S. et al. 2005. "Aroma Profiles of Pineapple Fruit (Ananas Comosus [L.] Merr.) and Pineapple Products." *LWT - Food Science and Technology* 38(3):263–74.

Erin Cooks. n.d. "The History of Pineapple." Retrieved January 27, 2017 (http://www.kitchenproject.com/history/Pineapple/).

Espinosa, Maita Eulalia Ávila et al. 2017. "Early Histological, Hormonal, and Molecular Changes during Pineapple (Ananas Comosus (L.) Merrill) Artificial Flowering Induction." *Journal of Plant Physiology* 209:11–19.

Farid Hossain, Md, Shaheen Akhtar, and Mustafa Anwar. 2015. "Nutritional Value and Medicinal Benefits of Pineapple." *International Journal of Nutrition and Food Sciences* 4(1):84–88.

Ferreira, F.R.; Cabral, J. R. S. 1998. *Melhoramento Genético Do Abacaxizeiro* [*Genetic Improvement of Pineapple*].

Ganem, Eduardo Luis de Oliveira. 2015. *A Cultura Do Abacaxizeiro* [*The Culture Of The Pineapple*]. Vitória da Conquista-Bahia.

Garcia, Wandreilla Moreira, Willian Krause, Isane Vera, and Rivanildo Dallacort. 2017. "Methods for Inoculation with Fusarium Guttiforme and Genetic Resistance of Pineapple (Ananas Comosus Var. Comosus)." *Revsita Caatinga*, 353–60.

Gomes, José Antônio et al. 2003. "Recomendações Técnicas Para a Cultura Do Abacaxizeiro." ["Technical Recommendations for the Culture of the Pineapple."] *DOCUMENTOS-Incaper* (122):27.

Hajar, Nadya et al. 2012. "Physicochemical Properties Analysis of Three Indexes Pineapple (Ananas Comosus) Peel Extract Variety N36." *APCBEE Procedia* 4:115–21.

Harold McGee. 1987. *On Food and Cooking (1984).*

Hazarika, Dipshika, Nabaneeta Gogoi, Seiko Jose, Robin Das, and Gautam Basu. 2017. "Exploration of Future Prospects of Indian Pineapple Leaf, an Agro Waste for Textile Application." *Journal of Cleaner Production* 141:580–86.

Hong, Keqian et al. 2013. "Quality Changes and Internal Browning Developments of Summer Pineapple Fruit during Storage at Different Temperatures." *Scientia Horticulturae* 151:68–74.

Jiang, Zhengdong, Hong Zheng, Nitin Mantri, Zhechen Qi, and Xiaodan Zhang. 2016. "Postharvest Biology and Technology Prediction of Relationship between Surface Area, Temperature, Storage Time and Ascorbic Acid Retention of Fresh-Cut Pineapple Using Adaptive Neuro-Fuzzy Inference System (ANFIS)." *Postharvest Biology and Technology* 113:1–7.

Júnior, J. Elias. 2009. "Integrated Management of Fusariosis in Pineapple Fields under Integrated Production System." 199–204.

Kaewtathip, Thipthida and Sanguansri Charoenrein. 2012. "Changes in Volatile Aroma Compounds of Pineapple (Ananas Comosus) during Freezing and Thawing." *International Journal of Food Science & Technology* 47(5):985–90.

Khayankarn, S., J. Uthaibutra, S. Setha, and K. Whangchai. 2013. "Using Electrolyzed Oxidizing Water Combined with an Ultrasonic Wave on the Postharvest Diseases Control of Pineapple Fruit Cv. ' Phu Lae.'" *Crop Protection* 54:43–47.

Li, Rui et al. 2016. "β-Cyclodextrin Assisted Two-Stage Foam Fractionation of Bromelain from the Crude Extract of Pineapple Peels." *Industrial Crops and Products* 94:233–39.

Li, Yun-he et al. 2016. "Molecular Cloning and Characterization of Four Genes Encoding Ethylene Receptors Associated with Pineapple (Ananas Comosus L.) Flowering." *Frontiers in Plant Science* 7(May):1–15.

Liu, C., Y. Liu, G. Yi, W. Li, and G. Zhang. 2011. "A Comparison of Aroma Components of Pineapple Fruits Ripened in Different Seasons." *African J.Agricult.Res.* 6(7):1771–78.

Liu, Jiazhu, Congcong He, Fei Shen, Kaili Zhang, and Shijiang Zhu. 2017. "The Crown Plays an Important Role in Maintaining Quality of Harvested Pineapple." *Postharvest Biology and Technology* 124:18–24.

Lu, Xinhua, Dequan Sun, Yunhe Li, Wenqi Shi, and Guangming Sun. 2011. "Pre- and Post-Harvest Salicylic Acid Treatments Alleviate Internal Browning and Maintain Quality of Winter Pineapple Fruit." *Scientia Horticulturae* 130(1):97–101.

Luengwilai, Kietsuda, Diane M. Beckles, and Jingtair Siriphanich. 2016. "Postharvest Internal Browning of Pineapple Fruit Originates at the Phloem." *Journal of Plant Physiology* 202:121–33.

Maeda, Alexandra Sanae et al. 2011. "Foliar Fertilization on Pineapple Quality and Yield." *Pesquisa Agropecuaria Tropical* [*Tropical Agriculture Research*] 41(2):248–53.

Manzoor, Zoya, Ali Nawaz, Hamid Mukhtar, and Ikram Haq. 2016. "Bromelain: Methods of Extraction, Purification and Therapeutic Applications." *Brazilian Archives of Biology and Technology* 59.

Matos, Aristoteles Pires de, Davi Theodoro Junghans, Eduardo Chumbinho de Andrade, and Paulo Ernesto Meissner Filho. 2011. "Impacto Potencial Das Mudanças Climáticas Sobre as Doenças Da Videira No Brasil." [Potential Impact of Climate Change on Vine Diseases in Brazil. In *Climate Change Impacts on important cultures in Brazil]* P. 331–56 (In Portuguese) in *Impactos das mudanças climáticas sobre doenças de importantes culturas no Brasil.*

Meletti, Laura Maria Molina, Aloísio Costa Sampaio, and Carlos Ruggiero. 2011. "Progress in the Brazilian Tropical Fruitculture." *Rev. Bras. Frutic* Special:73–75.

Mohan, Resmi, Venkatasubramanian Sivakumar, Thirumalaisamy Rangasamy, and Chellappa Muralidharan. 2016. "Optimisation of Bromelain Enzyme Extraction from Pineapple (Ananas Comosus) and Application in Process Industry." *American Journal of Biochemistry and Biotechnology* 12(3):188–95.

Montero-Calderón, Marta, María Alejandra Rojas-Graü, and Olga Martín-Belloso. 2010. "Aroma Profile and Volatiles Odor Activity Along Gold Cultivar Pineapple Flesh." *Journal of Food Science* 75(9):506–12.

Morais, Elisa Helena da Costa, Alessandra Aparecida Zinato Rodrigues, Maria Eliana Lopes Ribeiro de Queiroz, Antônio Augusto Neves, and

Paulo Henrique Damasceno Morais. 2014. "Determination of Thiamethoxam, Triadimenol and Deltamethrin in Pineapple Using SLE-LTP Extraction and Gas Chromatography." *Food Control* 42:9–17.

Motoagro. 2015. "Motoagro - Kit Acoplado de Poda, Adubação E Pulverização." *Motoagro Soluções Para O Agronegócios* [Motoagro - Pack Pruning, Fertilizing and Spraying Kit. *Motoagro Solutions for Agribusiness*]. Retrieved February 14, 2017 (http://www.motoagro.com.br/motoagro.html).

Nadzirah, K. Z., S. Zainal, A. Noriham, and I. Normah. 2013. "Efficacy of Selected Purification Techniques for Bromelain." *International Food Research Journal* 20(1):43–46.

Nogueira, Sônia R., Fábia S. O. Lima, Edivan M. Rocha, and Diego H. M. Araújo. 2014. "Fungicides in Fusariosis Pineapple Control in the State of Tocantins, Brazil." *Revista de Ciências Agrárias* 55 [*Journal of Agricultural Sciences]* 37(4):447–55.

Nor, M. Z. M., L. Ramchandran, M. Duke, and T. Vasiljevic. 2016. "Separation of Bromelain from Crude Pineapple Waste Mixture by a Two-Stage Ceramic Ultrafiltration Process." *Food and Bioproducts Processing* 98:142–50.

Offia-Olua, Blessing I. and O. A. Ekwunife. 2015. "Production and Evaluation of the Physico-Chemical and Sensory Qualities of Mixed Fruit Leather and Cakes Produced from Apple (Musa Pumila), Banana (Musa Sapientum), Pineapple (Ananas Comosus)." *Nigerian Food Journal* 33(1):22–28.

Pavan, Rajendra, Sapna Jain, Shraddha, and Ajay Kumar. 2012. "Properties and Therapeutic Application of Bromelain: A Review." *Biotechnology Research International* 2012:976203.

Perecin, T. N. et al. 2011. "Evaluation of the Effects of Gamma Radiation on Physical and Chemical Characterisitics of Pineapple (Ananas Comosus (L.) Meer) Cv. Smooth Cayenne Minimally Processed." *Progress in Nuclear Energy* 53(8):1145–47.

Raimbault, Astrid-Kim, Yasmine Zuily-Fodil, Alain Soler, and Maria H. Cruz de Carvalho. 2013. "A Novel Aspartic Acid Protease Gene from

Pineapple Fruit (Ananas Comosus): Cloning, Characterization and Relation to Postharvest Chilling Stress Resistance." *Journal of Plant Physiology* 170(17):1536–40.

Raimbault, Astrid Kim, Yasmine Zuily-Fodil, Alain Soler, Phillipe Mora, and Maria H. Cruz de Carvalho. 2013. "The Expression Patterns of Bromelain and AcCYS1 Correlate with Blackheart Resistance in Pineapple Fruits Submitted to Postharvest Chilling Stress." *Journal of Plant Physiology* 170(16):1442–46.

Razumiejczyk, Eugenia et al. 2016. "Crossmodal Interference between Language and Flavour." *Revista Latinoamericana de Psicología* 1–11.

Ren, Jing-Nan et al. 2015. "Characterisation of Free and Bound Volatile Compounds from Six Different Varieties of Citrus Fruits." *Food Chemistry* 185:25–32.

Rodin, J. O., Chester M. Himel, Robert M. Silverstein, Robert W. Leeper, and Willis A. Gortner. 1965. "Volatile Flavor and Aroma Components of Pineapple. L. Isolation and Tentative Identification of 2,5-Dimethyl-4-Hydroxy-3(2H)-Furanone." *Journal of Food Science* 30(2):280–85.

Rodríguez, Óscar, Wesley Gomes, Sueli Rodrigues, and Fabiano A. N. Fernandes. 2016. "Effect of Acoustically Assisted Treatments on Vitamins, Antioxidant Activity, Organic Acids and Drying Kinetics of Pineapple." *Ultrasonics Sonochemistry*.

Sampaio, R. A.; Araújo, W. F. 2001. "Importância de Cobertura Plástica Do Solo Sobre O Cultivo de Hortaliças." *Agropecuária Técnica* [Importance of Soil Plastic Coating on the Cultivation of Vegetables. *Technical Agriculture],* 22:1–12.

Samson, J. A. 1986. *Tropical Fruits*. 2nd ed. New York.: Longman Scientific & Technical.

Sanches, Nilton Fritzons and Aristóteles Pires de Matos. 2013. "Abacaxi, O Produtor Pergunta, a Embrapa Responde." [Pineapple, The Producer Asks, Embrapa Responds] 201.

Scherer, R. F., D. Olkoski, F. V. D. Souza, R. O. Nodari, and M. P. Guerra. 2015. *Gigante de Tarauacá: A Triploid Pineapple from Brazilian Amazonia*.

Selvarajah, S., A. D. Bauchot, and P. John. 2001. "Internal Browning in Cold-Stored Pineapples Is Suppressed by a Postharvest Application of 1-Methylcyclopropene." *Postharvest Biology and Technology* 23(2):167–70.

Silva, Keila S., Milena A. Fernandes, and Maria A. Mauro. 2014. "Effect of Calcium on the Osmotic Dehydration Kinetics and Quality of Pineapple." *Journal of Food Engineering* 134:37–44.

Silva, Websten Cesário da. 2007. *Sistema de Produção Para a Cultura Do Abacaxi No Estado de Rondônia* [*Production System for Pineapple Culture in the State Of Rondônia*] Porto Velho- RO.

Soares, A. G., L. C. Trugo, N. Botrel, and L. F.da S. Souza. 2005. "Reduction of Internal Browning of Pineapple Fruit (Ananas Comusus L.) by Preharvest Soil Application of Potassium." *Postharvest Biology and Technology* 35(2):201–7.

Soares, A. G., L. C. Trugo, N. B. Gonçalves, S. A. Cenci, and M. J.de O. Fonseca. 2005. *Uso de Diferentes Fibras de Micro-Extracao Em Fase Solida Para Identificacao de Compostos Volateis Em Abacaxi Cultivar Smooth Cayenne*. Rio de Janeiro. [*Use of Different Solid-Phase Micro-Extraction Fibers for Identification of Volatile Compounds in Pineapple Cultivate Smooth Cayenne*. Rio de Janeiro.]

Soares, Paulo A. G., Antônio F. M. Vaz, Maria T. S. Correia, Adalberto Pessoa, and Maria G. Carneiro-da-Cunha. 2012. "Purification of Bromelain from Pineapple Wastes by Ethanol Precipitation." *Separation and Purification Technology* 98:389–95.

De Souza, Jorge T., Rafael O. Trocoli, and Fernando P. Monteiro. 2016. "Plants from the Caatinga Biome Harbor Endophytic Trichoderma Species Active in the Biocontrol of Pineapple Fusariosis." *Biological Control* 94:25–32.

Spironello, A.; Siqueira, W.J.; Martins, A.L.M.; Usberti Filho, J.A.; Carvalho, C.R.L.; Bettiol Neto, J.E.; Sigrist, J.M.M.; Ferrari, J.T.; Louzeiro, I.M. 2011. "Avaliação Do Híbrido de Abacaxizeiro IAC Fantástico Visando À Indicação de Cultivo". P. 4 in *Anais Bauru*. [Evaluation of the Pineapple Hybrid IAC Fantastic for the Indication of Cultivation P. 4 In *Anais Bauru*].

Steingass, Christof B., Tara Grauwet, and Reinhold Carle. 2014. "Influence of Harvest Maturity and Fruit Logistics on Pineapple (Ananas Comosus [L.] Merr.) Volatiles Assessed by Headspace Solid Phase Microextraction and Gas Chromatography–mass Spectrometry (HS-SPME-GC/MS)." *Food Chemistry* 150:382–91.

Steingass, Christof B., Johannes Langen, Reinhold Carle, and Hans Georg Schmarr. 2014. "Authentication of Pineapple (Ananas Comosus [L.] Merr.) Fruit Maturity Stages by Quantitative Analysis of γ- and δ-Lactones Using Headspace Solid-Phase Microextraction and Chirospecific Gas Chromatography-Selected Ion Monitoring Mass Spectrometry (HS-SPME." *Food Chemistry* 168:496–503.

Tokitomo, Yukiko, Martin Steinhaus, Andrea Büttner, and Peter Schieberle. 2005. "Odor-Active Constituents in Fresh Pineapple (Ananas Comosus [L.] Merr.) by Quantitative and Sensory Evaluation." *Bioscience, Biotechnology and Biochemistry* 69(7):1323–30.

Uchôa, Cleilson do Nascimento et al. 2014. "Acibenzolar-S-Methyl and Silicium as a Inductor of Resistance to Black Sigatoka of Banana Cultivar Grand Naine (AAA)." *Revista Agrarian* 7(24):189–96.

Vartiainen, Jenni, M. Aksela, and A. Hopia. 2013. "Introduction to Molecular Gastronomy and to Its Applications in Science Education." *Lumat* 1(2):143–50.

Ventura, J. A., H. Costa, J. R. S. Cabral, and A. P. De Matos. 2009. "'Vitória': New Pineapple Cultivar Resistant to Fusariosis." *Acta Horticulturae* 822:51–55.

Vilanova Neta, Jaci Lima et al. 2012. "Bromelain Enzyme from Pineapple: In Vitro Activity Study under Different Micropropagation Conditions." *Applied Biochemistry and Biotechnology* 168(2):234–46.

Wei, Chang Bin et al. 2011. "Characteristic Aroma Compounds from Different Pineapple Parts." *Molecules* 16(6):5104–12.

Ye, Liqin et al. 2017. "Evaluation of Volatile Compounds from Chinese Dwarf Cherry (Cerasus Humilis (Bge.) Sok.) Germplasms by Headspacc Solid-Phase Microextraction and Gas Chromatography–mass Spectrometry." *Food Chemistry* 217:389–97.

Yeoh, Wei Keat and Asgar Ali. 2017. "Ultrasound Treatment on Phenolic Metabolism and Antioxidant Capacity of Fresh-Cut Pineapple during Cold Storage." *Food Chemistry* 216:247–53.

Zhang, Bao-Yu et al. 2013. "Effect of Atmospheres Combining High Oxygen and Carbon Dioxide Levels on Microbial Spoilage and Sensory Quality of Fresh-Cut Pineapple." *Postharvest Biology and Technology* 86:73–84.

Zheng, Liang Yong et al. 2012. "Aroma Volatile Compounds from Two Fresh Pineapple Varieties in China." *International Journal of Molecular Sciences* 13(6):7383–92.

Zhou, Yuchan, Janelle M. Dahler, Steven J. Underhill, and Ron B. Wills. 2003. "Enzymes Associated with Blackheart Development in Pineapple Fruit." *Food Chemistry* 80(4):565–72.

In: Tropical Fruits
Editors: C. Stewart Bogsan et al.
ISBN: 978-1-53612-885-7

Chapter 9

GENE SILENCING BY MICRORNA IN PINEAPPLE: DISCOVERY, INVOLVEMENT IN THE CONTROL OF FRUIT DEVELOPMENT AND ITS APPLICATION AS ARTIFICIAL GENE REGULATORS

***Noor Hydayaty Md. Yusuf*[1], *Mariam Abd. Latip*[2] *and Vijay Kumar Subbiah*[1,*]**

[1]Biotechnology Research Institute, University Malaysia Sabah, Kota Kinabalu, Sabah, Malaysia

[2]Faculty of Science and Natural Resources, University Malaysia Sabah, Kota Kinabalu, Sabah, Malaysia

ABSTRACT

MicroRNAs (miRNAs) are a class of small RNAs, usually 19–24 nt in length, which are found endogenously within the cell and do not code

---

[*] Corresponding Author Email: vijay@ums.edu.my.

for any protein. However, they participate in regulating the level of mRNA transcripts through cleavage or translational inhibition, creating an effect called gene silencing. MicroRNAs have been shown to be essential for major biological and physiological development in plants, including pineapple. A hundred and fifty-three miRNAs, regulating many aspects of plant growth, have been described in pineapple to date. Reports of the existence of this natural gene silencing system have led to the development of a similar system at a synthetic level. Artificial microRNA (amiRNA) is a unique custom-designed molecule of RNA, approximately 21 nt in length, with the sole function of silencing the expression of its target gene by mimicking the action of miRNA in the RNA interference (RNAi) pathway. In pineapple, target genes have not only been silenced using this technique but the silencing has occurred in a specific manner, i.e., the target genes were silenced without affecting the expression of other, unrelated genes. This technique has addressed the limitations of conventional breeding techniques, as amiRNA silencing can be performed rapidly and is time-consuming and occurs in a specific manner.

Keywords: MicroRNA, artificial microRNA, MD2 pineapple, gene silencing

INTRODUCTION

Crops, by definition, are plants grown to be harvested for agricultural use. The agricultural industry is dependent on human need. Therefore, agricultural crops are essentially plants with economically desirable traits. The potential of a plant (to be used as a crop) is identified by its phenotype, which is encoded by specific genes in the plant's genome. Theoretically, one gene codes for one protein, while a combination of one or more proteins results in some phenotype(s) [1]. Recognising the importance of this process, significant effort to identify the presence and function of various genes in a plant have been described and are ongoing. Knowledge of gene(s) that contribute to a phenotype (both desirable and undesirable) have facilitated breeders in the development of efficient breeding programs.

The availability of different techniques and technologies developed from the field of biotechnology have benefitted this effort, and studies on both types of genetic material, DNA and RNA, can be performed. As both DNA and RNA affect gene expression (as well as the expression of a phenotype), the study of the interaction between the two gives a broader view of how a gene is regulated. As such, some protein-coding genes are regulated by small RNAs, called microRNAs (miRNAs) [2, 3]. MicroRNAs function to regulate a gene by inhibiting its expression. This is termed gene silencing by a miRNA, a mechanism which occurs naturally in all organisms. This chapter explains how miRNAs were identified in pineapple and how their expression may regulate the protein-coding genes involved in fruit development. Also, the utilisation of this theory in the production of synthetic miRNAs, called artificial miRNAs (amiRNA), for custom gene silencing in pineapple is described.

The Concept of Gene Silencing by MicroRNA

Genes are essentially classified as segments of DNA (on a specific site on a chromosome) that are responsible for the physical and inheritable characteristics or phenotype of an organism. Generally, a gene will be transcribed to a mRNA transcript, before being translated into a protein. This process is called the central dogma of molecular biology. The success of a gene to produce specific protein through this pathway (DNA - mRNA - protein) is called gene expression. However, not all mRNAs will be translated into proteins. Some will have miRNA bound to it, and prevented from being translated into proteins [4, 5].

MicroRNAs (miRNA) are a type of small RNA (~21nt) which is processed from a longer sequence of a secondary structure called the primary microRNA (pri-miRNA). Since this pri-miRNA sequence is derived from long transcripts, it forms a rather complex secondary structure containing several branches (each with a stem and a loop). The approximately 21 nt miRNA sequence is located at the stem of one of these branches. The branch-containing miRNA, which is called the precursor miRNA (pre-miRNA or precursor), is released from the pri-miRNA through cleavage by the RNase III containing protein, DICER-LIKE1 protein (DCL1). The next step in miRNA biogenesis involves the release of miRNA from the precursor.

Similarly, the sequence is released from the precursor by the DCL1 proteins through the cleavage process. The miRNA released is in duplex form, and the sense strand is called miRNA while the passenger strand is called miRNA*. The miRNA sequence of the duplex binds to a ribonucleoprotein complex that contains ARGONAUTE1 (AGO1) protein and is called RNA Induced Silencing Complex, or RISC. Since miRNAs are small and single stranded, it can bind complementarily to other single-stranded sequences such as the mRNA transcripts. When this occurs, the AGO1 protein performs cleavageof the transcript, distrupting the translation of mRNA, and no protein is produced.

Some miRNAs are unable to perfectly adhere to the transcript. If this occurs, AGO1 cannot execute the cleavage process. However, the RISC complex will remain attached to the transcript and repress the translation of the mRNA transcript.

DISCOVERY OF MICRORNA IN PINEAPPLE

The presence of microRNA in an organism can be identified by various methods. In pineapple, conventional cloning and sequencing of sRNA were among the first methods used [6, 7].

Eighty-three conserved miRNAs belonging to twelve families have been identified. The number of identified miRNAs in pineapple subsequently increased (153 miRNAs belonging to 41 families) when next generation sequencing was applied [8]. The abundancy of each miRNA are different, with the highest recorded from miR159 followed by miR319, miR535, miR396, and miR156, miR166, miR530, miR162, miR157, miR827, miR529, miR528, miR164, miR172, miR171, miR167, miR444, miR894, miR169, miR160, miR394, miR399, miR393, miR896, miR845, miR397, miR391, miR1511, miR2916, miR1432, miR168, miR850, miR2677, miR1211, miR893, miR1873, miR165, miR395, miR1088, miR1426, and miR390 (Figure 1).

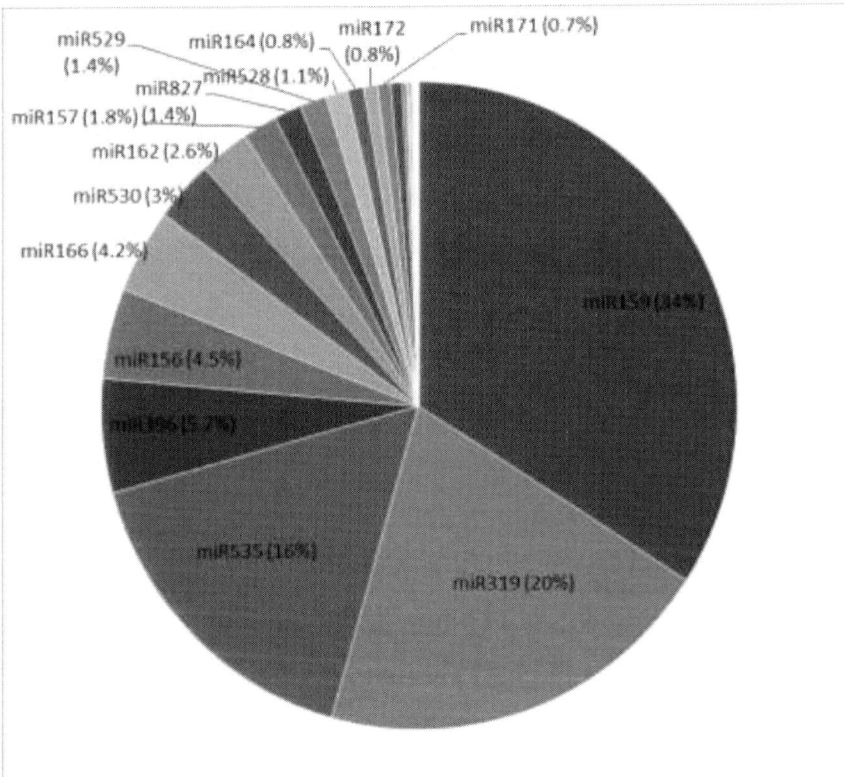

Figure 1. Fifteen highly abundant miRNA in pineapple. The most abundant is miR159 (representing 34% of the miRNA population in pineapple) followed by miR319, miR535, miR396, miR156, miR166, miR530.

Function and Expression of Microrna in Pineapple

Many miRNAs have been identified, although most have not been functionally profiled. Functional profiling of miRNAs requires the creation of an overexpression or lost-of-function mutant, a relatively time-consuming process. Thus, alternatively, functionality has instead been predicted through bioinformatics analyses. Since miRNAs exert their effects by binding to their target transcript, mRNAs with a binding site (a sequence which complements the miRNA) are considered as the target. In pineapple, the availability of significant numbers of mRNA transcripts has enabled target prediction [7-10]. Potential targets have been assigned to 17 miRNAs of different families in pineapple, with most of the target being transcription factors (Table 1).

The discovery of miRNAs and its target transcript indicates the presence of a gene regulation mechanism through silencing in pineapple. During fruit development alone, at least 32 genes may be regulated by miRNAs, as 32 miRNAs (out of 41 miRNA families) have been found to be expressed during this process [8].

Table 1. Pineapple miRNAs and their predicted target transcripts and functions

miRNA	Conserved target gene	Predicted function of target gene
miR156	SQUAMOSA transcription factor	Phase transition
miR159	MYB transcription factor	Gibberellic and abscisic acid biosynthesis
miR162	DCL1 Dicer-like 1	miRNA biogenesis
miR164	NAC transcription factor	Meristem boundary
miR165	HD-ZIP III transcription factor	Cell multiplication
miR167	ARF transcription factor	Auxin biosynthesis
miR168	AGO1 ARGONAUTE1 protein	miRNA biogenesis
miR171	SCL transcription factor	Cell division
miR172	AP2 transcription factor	Flowering
miR319	TCP transcription factor	Jasmonic acid biosynthesis
miR393	TIR1 transcription factor	Auxin biosynthesis
miR395	SULTR	Sulphate uptake and homeostasis
miR397	Laccase	Copper uptake and homeostasis
miR396	GRF transcription factor	Cell proliferation
miR399	PO4	Phosphate uptake and homeostasis
miR444	MADS-box	Cell differentiation
miR535	XET enzyme	Cell wall developement

These miRNAs were differentially expressed (upregulated or downregulated) at different stages of fruit development (mature and ripe pineapple fruit) when profiled using stem-loop Reverse Transcription-quantitative Polymerase Chain Reaction (RT-qPCR). These miRNAs were highly upregulated or downregulated by minimum of 0.2 fold to 3.5 fold (Figures 2, 3, 4, 5).

As the interaction between miRNAs and the target mRNAs occur in reverse order, i.e., high expression of miRNA leads to low expression of mRNA, thus, upregulation of miRNA is projected to downregulate the expression of the predicted target transcript, and vice versa. Based on this differential expression pattern, miRNAs have been predicted to be involved in at least four major events throughout the fruit development process, including fruit development, phase and stage transition, hormonal signalling, nutrient uptake and homeostasis, and miRNA homeostasis.

Fruit Development of Pineapple Regulated by miRNAs

Five miRNAs are postulated to be involved in fruit development of pineapple and the model is as shown in Figure 2. Fruit development involves cell division, multiplication and proliferation activities. These three processes are thought to be controlled by miR165, miR396 and miR444. Decreased expression of miR165 and miR396 during fruit development appears to result in high expression of the target genes, HD-ZIP III and GRF transcription factor, respectively, thus, stimulating cell division, multiplication and proliferation. In plants, cell proliferation is the first process to be observed throughout the developing organ, and starts at the tip of a particular organ [11, 12]. This process is followed by mitotic arrest, which moves toward the base of the organ where cell division regulates organ growth, enlargement and size [12]. Next, low expression of miR444 in the early stage of fruit development promotes cell differentiation, as the target gene (MADS-box) is expressed. MADS-box proteins belong to a class of transcription factors and are crucial in organ and cell differentiation [13, 14].

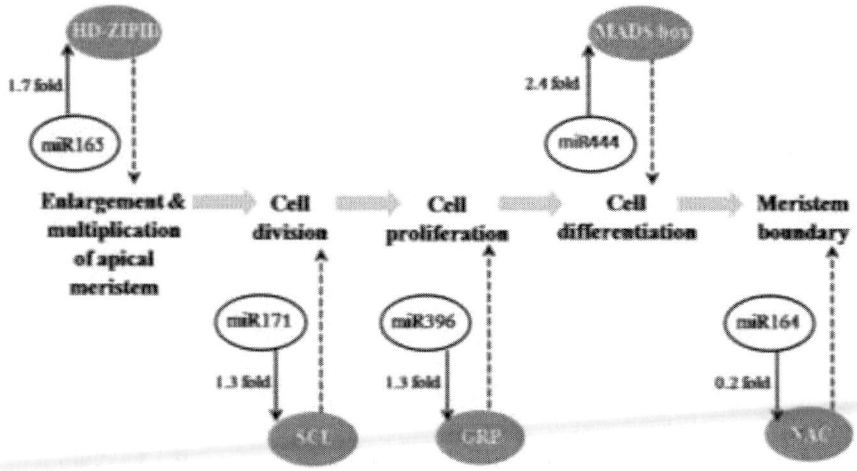

Figure 2. A model of pineapple fruit development networks regulated by miRNAs based on stem-loop RT-qPCR analysis. Differential expression between two stages of fruit (mature and ripe pineapple) is as indicated by fold differences.

Finally, at the end of the fruit development process, the expression of miR164 is downregulated to release the NAC gene which determines the fruit border and size. In plants, boundary establishment is controlled by NAC-domain genes including CUP-SHAPED COTYLEDON1 (CUC1) and CUC2 [15]. NAC has also been reported to regulate the separation of the organs by restricting cell proliferation [16].

Phase and Stage Transition of Pineapple Fruit Regulated by miRNAs

Phase transition involves the change from flowering into fruit maturation, softening and finally senescence. A model of pineapple phase and stage transition throughout fruit development is shown in Figure 3.

First, the silencing of AP2 by upregulation of miR172 postulated to suppresses its interaction with the TOE pathway and triggers the transition from flowering to fruit development [17]. Next, the upregulation of miR156 indicates its involvement in phase transition during fruit development. MiR156 is established as a controller of phase transition in plants [17]. Increases in miR535 during fruit development is believed to silence the expression of XET and trigger cell wall degradation and loosening, an event necessary for fruit ripening. This observation supports the characteristics of the pineapple fruit, implying that throughout pineapple fruit maturation, its size increases as the cell wall thins [18]. Pineapple miR319 is believed to target TCP transcription factor genes involved in jasmonic acid (JA) biosynthesis, which has been shown to strongly promote senescence [19, 20].

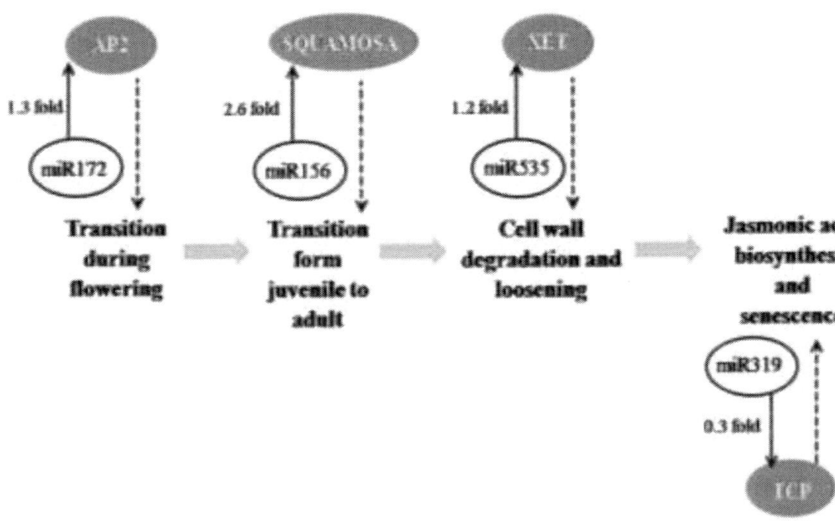

Figure 3. Model of pineapple phase and stage transition throughout fruit development by miRNAs based on stem-loop RT-qPCR analysis. Differential expression between two stages of fruit (mature and ripe pineapple) is indicated by fold differences.

Nutrient Uptake and Homeostasis Regulated by miRNAs

Three miRNAs are postulated to be involved in nutrient uptake and homeostasis in pineapple (Figure 4). As for miR395, the sulphate transporter (SULTR) transcripts targeted are crucial in plant development. Since most sulphur sre derived from inorganic sulphate (SO_4^{2}), where it is found in the soil, SULTR transcripts are responsible for the initial uptake of sulphate at the root surface and later mediate the translocation of internal sulphate within plants [21].

MiR399 are complementary to target transcripts coding for ubiquitin-conjugating enzymes, which are known to regulate phosphate assimilation in plants.

In *Arabidopsis*, the expression of miR397 in environments lacking copper indicate that miR397 regulates the laccase copper protein family [22, 23], a function thought to be similar in pineapple.

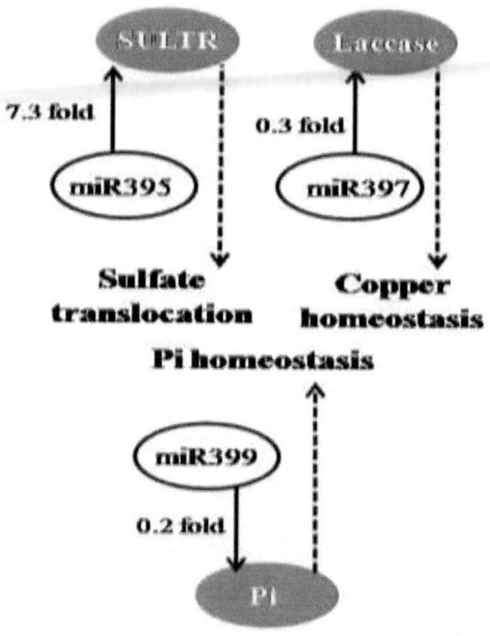

Figure 4. Model of pineapple nutrient uptake and homeostasis regulated by miRNAs postulated based on stem-loop RT-qPCR analysis. Differential expression between two stages of fruit (mature and ripe pineapple) is indicated by fold differences.

Hormone Signalling in Pineapple Regulated by miRNAs

Hormone signalling in plant is reported to be controlled by three miRNAs, and these are postulated to play their same roles in pineapple (Figure 5). It has been suggested that MYB transcripts are highly expressed in response to the absence of miR159 during ripening. MYB regulates gibberellin (GA) and abscisic acid (ABA) signalling in plants [24, 25].

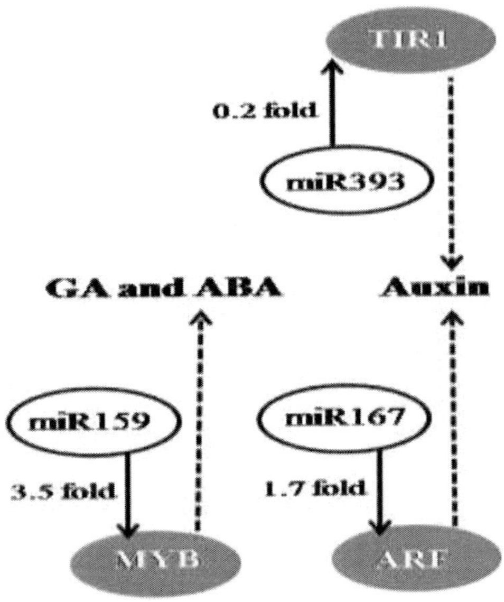

Figure 5. Model of pineapple hormone signalling regulated by miRNAs postulated based on stem-loop RT-qPCR analysis. Differential expression between two stages of fruit is indicated by fold differences.

Two additional pineapple miRNAs thought to regulate the expression of auxin are miR167 and miR393. The target prediction of pineapple miR167 correlates with mRNA transcripts coding for auxin response factor (ARF). Cleavage of TIR1 by miR393 in pineapple believed to directly affects the expression of AUX/IAA proteins for proteolysis, an event necessary for auxin-induced growth processes [26, 27].

The Concept of Gene Silencing by Artificial MicroRNA

The biogenesis of miRNA explains the mechanism whereby the *MIRNA* coding genc is processed to produce miRNA in order to negatively regulate gene expression within an organism. Several techniques have been developed based on the concept of gene silencing by endogenous miRNA, and one of it is the artificial microRNA (amiRNA). AmiRNA is a genetic-based technology developed to mimic gene silencing by miRNA [28]. Its difference, however, also represents an advantage over the use of miRNA, whereby it can be custom-designed to silence a specific target gene within an organism. An amiRNA is a ~21 nt oligonucleotide with a sequence that is complimentary to the targeted mRNA sequence. It has been reported that, when the sequence was inserted into the endogenous pre-miRNA (backbone), thus replacing the natural ~21 nt miRNA, it was able to function normally (i.e., able to produce amiRNA). AmiRNA then binds to the target mRNA and silences it. The establishment of amiRNA technique holds significant potential for the development of new varieties in pineapple through large-scale yet specific gene silencing. In this case, silencing each gene on one particular pathway (individually or the whole family) [28, 29]..

ARTIFICIAL MICRORNA FOR CUSTOM GENE SILENCING IN PINEAPPLE

AmiRNA has been established and widely used for the creation of loss-of-function mutants/lines in plants and commercial crops such as *Oryza sativa* and *Zea mays* [29, 30]. However, since pre-miRNA is known to be species specific, precursors used as amiRNA backbones in these plants may or may not be compatible with other plant species. In pineapple, an amiRNA system for gene silencing has been established by utilizing three endogenous pre-miRNAs, namely pre-miR156, pre-miR399 and pre-miR2673 [8]. Artificial microRNAs targeting *MIR535* genes (miR535b and miR535m) which are postulated to be involved in cell wall development were designed and inserted into pineapple pre-miRNA (Figure 6) through *Agrobacterium*-mediated transformation [31]. The expression of amiRNA and the target gene in these lines were profiled using stem-loop RT-qPCR (unpublished data).

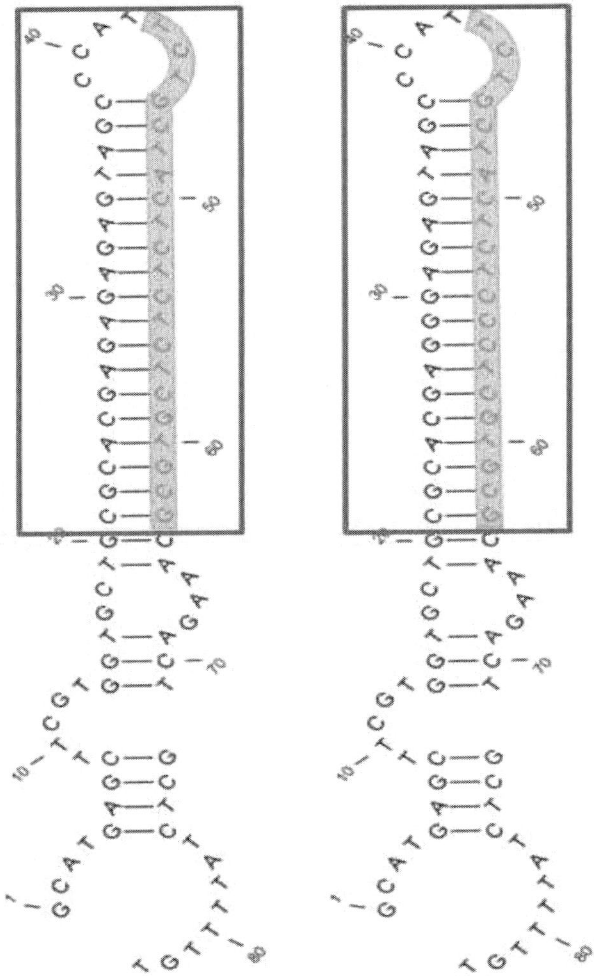

Figure 6. The pre-miR399 of pineapple inserted with two amiRNAs, replacing the mature miRNA region (highligted). These amiRNAs target two *MIR535* members i.e., miR535b and miR535m. Numbers linked to the nucleotide represents their position from the 5’ end of pre-miR399.

The pineapple precursors were tested for their efficiency, which was measured through their ability to first express high levels of amiRNA and second to silence the target gene to a high degree (Figure 7). Two precursors (pre-miR156 and pre-miR319) were categorized as highly efficient backbones, because they revealed a high accumulation of amiRNA and a strong silencing effect. These newly designed amiRNAs were successful in silencing the endogenous non-protein-coding *MIR535* family (miR535b and miR535m) in pineapple.

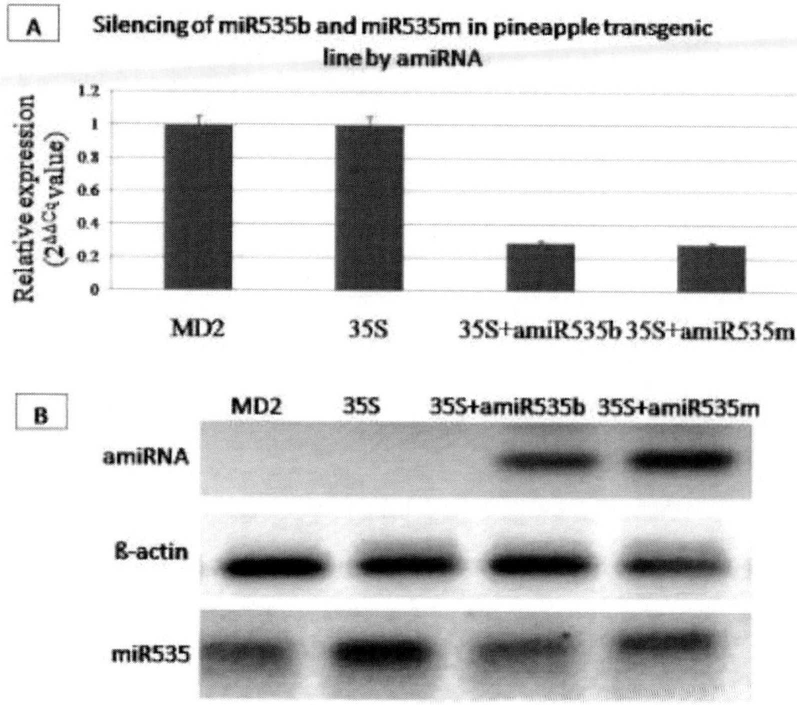

Figure 7. Silencing of miR535b and miR535m by amiRNA in pineapple. (A) Relative expression of miR535b and miR535m obtained through stem-loop RT-qPCR comparing control callus (MD2 and vector control) and transgenic callus (transformed with pre-amiRNA expression vector); (B) Amplification of amiRNA, β-action (RT-qPCR reference gene) and miR535 gene comparing control callus (wild and vector control) and transgenic callus (transformed with pre-amiRNA expression vector).

Conclusion

The numerous miRNAs found expressed in pineapple may hold the key to the control of various economically desired traits. However, these miRNA are yet to be functionally profiled and characterised. Thus, miRNA-target-phenotype relationship studies are needed. With the combination of highly efficient pre-amiRNA expression vectors and transformation techniques, large-scale functional profiling endeavours can be carried out. The functions of thousands of genes in pineapple can be profiled by the creation of loss-of-function mutants on a large scale using currently available high-throughput technology. The impact of key genes or gene families that act as a 'switch off button' for a particular metabolic pathway can now be identified with miRNAs, leading to a more focused breeding program.

References

[1] Klug, W. S., Cummings, M. R. and Spencer, C. A. (1997). *Concepts of Genetics* (8th edition). New Jersey, United States of America: Pearson Education. Inc.

[2] Ambros, V. (1989). A hierarchy of regulatory genes controls a larva-to-adult developmental switch in *C. elegans*. *Cell,* 57:49-57.

[3] Lee, R. C., Feinbaum, R. L. and Ambros, V. (1993). The *C. elegans* heterochronic gene *lin-4* encodes small RNAs with antisense complementary to *lin-14*. *Cell,* 75:843-854.

[4] Carthew, R. W. and Sontheimer, E. J. (2009). Origins and mechanisms of miRNAs and siRNAs. *Cell,* 136:642-655.

[5] Kurihara, Y., Takashi, Y. and Watanabe, Y. (2006). The interaction between DCL1 and HYL1 is important for efficient and precise processing of pri-miRNA in plant microRNA biogenesis. *RNA,* 12:206-212.

[6] Yew, C. W. and Kumar, S. V. (2012). Isolation ad cloning of microRNAs from recalcitrant plant tissues with small amounts of total RNA: a step-by step approach. *Molecular Biology Reports,* 39:1783-1790.

[7] Yew, C. W. and Kumar, S. V. (2011). MicroRNA regulates gene expression during fruit development in pineapple. *Acta Horticulturae,* 902:177-184.

[8] Yusuf, N. H. M., Ong, W. D., Redwan, R. M., Latip, M. A. and Kumar, S. V. (2015). Discovery or precursor and mature microRNAs and their putative gane targets using high-throughput sequencing in pineapple (*Ananas comosus* var. *comosus*). *Gene,* 571:71-80.

[9] Yusuf, N. H. M. and Kumar, S. V. (2011). Identification and characterization of differentially expressed microRNAs during fruit ripening in Pineapple (*Ananascomosus* var. *comosus*). *Acta horticulturae,* 902:123-131.

[10] Ong, W. D., Voo, L. Y. C. and Kumar, S. V. (2012). De Novo assembly, characterization and functional annotation of pineapple fruit transcriptome through massively parallel sequencing. *Plos One,* 7:1-11.

[11] Jung, J. H. and Park, C. M. (2007). MiR166/165 genes exhibit dynamic expression patterns in regulating shoot apical meristem and floral development in *Arabidopsis*. *Planta,* 225:1327-1338.

[12] Piazza, P., Jasinski, S. and Tsiantis, M. (2005). Evolution of leaf developmental mechanism. *New Phytologist,* 167:693-710.

[13] Gaffe, J., Lemercier, C., Alcaraz, J. P. and Kuntz, M. (2011). Identification of three tomato flower and fruit MADS-box proteins with a putative histone deacetylase binding domain. *Gene,* 471:19-26.

[14] Gu, Q., Ferrandiz, C., Yanofsky, M. F. and Martienssen, R. (1998). The FRUITFULL MADS-box gene mediates cell differentiation during *Arabidopsis* fruit development. *Development,* 125:1509-1517.

[15] Laufs, P., Peaucelle, A., Morin, H. and Traas, J. (2004). MicroRNA regulation of the CUC genes is required for boundary size control in *Arabidopsis* meristems. *Development,* 131:4311-4322.

[16] Mallory, A. C., Dugas, D. V., Bartel, D. P. and Bartel, B. (2004). MicroRNA regulation of NAC-domain targets is required for proper formation and separation of adjacent embryonic, vegetative, and floral organs. *Current Biology,* 14:1035-1046.

[17] Wu, G. and Poething, R. S., (2006). Temporal regulation of shoot development in *Arabidopsis thaliana* by miR156 and its target SPL3. *Development,* 111:3539-3547.

[18] Bartholomew, D. P., Paull, R. E. and Rohrbach, K. G. (2003). *The pineapple: botany, production and uses*. Bartholomew, D. P., Paull, R. E., Rohrbach, K. G. (eds). Wallingford, United Kingdom: CABI Publishing.

[19] Schommer, C., Palatnik, J. F., Aggarwal, P., Chételat, A., Cubas, P., Farmer, E. E., Nath, U. and Weigel, D. (2008). Control of jasmonate biosynthesis and senescence by miR319 targets. *Plos Biology,* 6:1991-2001.

[20] Salisbury, F. B. and Ross, C. W. (1992). *Plantphysiology*. California; Wadsworth, Inc.

[21] Chiou, T. J. (2007). The role of microRNAs in sensing nutrient stress. *Plant, Cell and Environmnet,* 30:323-332.

[22] Abdel-Ghany, S. E. and Pilon, M. (2008). MicroRNA-mediated systemic down-regulation of copper protein expression in response to low copper availability in *Arabidopsis*. *Journal of Biological Chemistry,* 282:15932-15945.

[23] Yamasaki, H., hayashi, M., Fukazawa, M., Kobayashi, Y. and Shikanai, T. (2009). SQUAMOSA promoter binding protein-like 7 is a central regulator for copper homeostasis in *Arabidopsis*. *Plant Cell,* 21:347-361.

[24] Gubler, F., Kalla, R., Roberts, J. K. and Jacobsen, J. V. (1995). Gibberellin-regulated expression of a *myb* gene in Barley Aleurone cells: evidence of *myb* transactivation of a high a-amilase gene promoter. *The Plant Cell,* 7:1879-1891.

[25] Mishra, N. S. and Mukherjee, S. K. (2007). A peep into the plant miRNA world. *The Open Plant Science Journal,* 1:1-7.

[26] Jones-Rhoades, M. and Bartel, D. P. (2004). Computational identification of plant microRNAs and their targets, including a stress-induced miRNA. *Molecular Cell,* 14:787-799.

[27] Parry, G., Calderon-Villalobos, L. I., Prigge, M., Peret, B., Dharmasiri, S., Itoh, H., Lechner, E., Gary, W. M., Bennett. and M., Estelle, M. (2009). Complex regulation of the TIR1/AFB family of auxin receptors. *Proceedings of the National Academy of Sciences of the United States of America,* 16:22540-22545.

[28] Schwab, R., Ossowski, S., Riester, M., Wartmann, N. and Weigel, D. (2006). Highly specific gene silencing by artificial microRNAs in *Arabidopsis*. *The Plant Cell,* 18:1121-1133.

[29] Wartmann, N., Cheh, H., Ossowski, S., Weigel, D. and Herve, P. (2008). Highly specific gene silencing by artificial miRNAs in rice. *Plos One,* 3:1-10.

[30] Meng, X., Muszynski, M. G. and Danilevskaya, O. N. (2011). The *FT*-like *ZCN8* gene functions as a floral activator and is involved in photoperiod sensitivity in maize. *The Plant Cell,* 23:942-960.

[31] Yusuf, N. H. M., Latip, M. A. and Kumar, S. V. (2016). Efficient protocol for plant regeneration and from callus cultures and *Agrobacterium* transformation of MD2 pineapple. *Transactions of Persatuan Genetik Malaysia,* 3:137-141.

Editors Contact Information

Dr. Cristina Stewart Bogsan
Department of Biochemical-Pharmaceutical Technology
Faculty of Pharmaceutical Sciences
University of São Paulo
05508-000, São Paulo, Brazil
Email: cris.bogsan@usp.br

Svetoslav Dimitrov Todorov
Food Research Center (FoRC),
Department of Food and Experimental Nutrition,
Faculty of Pharmaceutical Sciences, University of São Paulo,
580, Professor Lineu Prestes,
13B, São Paulo, SP, 05508-000, Brazil.
Email: slavi310570@abv.bg

INDEX

A

C

E

F